Workplace

Workplace Bullying

What we know, who is to blame,
and what can we do?

**Charlotte Rayner, Helge Hoel
and Cary L. Cooper**

London and New York

First published 2002
by Taylor & Francis
11 New Fetter Lane, London EC4P 4EE

Simultaneously published in the USA and Canada
by Taylor & Francis Inc.
29 West 35th Street, New York, NY 10001

Taylor & Francis is an imprint of the Taylor & Francis Group

Typeset in Sabon by Exe Valley Dataset Ltd, Exeter
Printed and bound in Great Britain by
The Cromwell Press, Trowbridge, Wiltshire

British Library Cataloguing in Publication Data
A catalogue record for this book is available
from the British Library

Library of Congress Cataloging in Publication Data
Rayner, Charlotte.
    Workplace bullying: what we know, who is to blame, and what
can we do? / Charlotte Rayner, Helge Hoel, Cary L. Cooper
       p. cm.
    Includes bibliographical references and index.
    1. Bullying in the workplace. 2. Bullying in the workplace–
Prevention. 3. Harassment. 4. Intimidation. I. Hoel, Helge.
II. Cooper, Cary L. III. Title.

HF5549.5.E43 R39 2001
658.3′145–dc21   2001018834

ISBN 0–415–24062–X (hbk)
ISBN 0–415–24063–8 (pbk)

Contents

Figures

Tables

Foreword

Bullying at work as an issue may have only come to prominence in the 1990s. However, it has existed in many organisations, including for instance the police service, for years. Since the incidence of bullying within the workplace was highlighted through the campaign by the indefatigable Andrea Adams before her untimely death in 1995, this problem has been real and debilitating.

Joining the Campaign Against Bullying at Work was a direct result of a direction from beyond the grave! I had come to know Andrea Adams not long before my daughter Suzy was murdered. Andrea became a great friend whose work on bullying I much admired. When she became ill I asked how I could help progress her work. The answer came in a letter, followed up by a note in *The Times* (alongside the notice of her death) that 'The Suzy Lamplugh Trust are making sure that her work would flourish'. We had no choice. However, we have never regretted her delegation. Nor do we ever forget the debt we owe her, and we pay tribute to the way she raised the public profile of this scourge at work.

Workplace bullying constitutes unwanted, offensive, humiliating, under-mining behaviour towards an individual or groups of employees. Such per-sistently malicious attacks on personal or professional performance are typically unpredictable, irrational and often unfair. This abuse of power or position can cause such chronic stress and anxiety that people gradually lose belief in themselves, suffering physical ill health and mental distress as a result.

Workplace bullying affects working conditions, health and safety, domestic life and the right of all to equal opportunity and treatment. Workplace bullying is a separate issue from sexual or racial harassment. It is a gradually wearing-down process which makes individuals feel demeaned and inadequate, that they can never get anything right and that they are hopeless not only within their work environment but also in their domestic life. In many instances bullying can be very difficult to detect. It often takes place where there are no witnesses. It can be subtle and devious and often it is difficult for those on the receiving end to confront their perpetrator.

The Suzy Lamplugh Trust is concerned with problems of aggression and violence at work. Looking at the consequences of bullying it became obvious that this problem could be considered an insidious and worrying personal safety issue. Research began to reveal that the stress created could lead to incidents, accidents and acts of carelessness or even self-damage. The Suzy Lamplugh Trust is the leading authority on personal safety. Its role is to minimise aggression in all its forms – physical, verbal and psychological. Whatever its nature, aggression is damaging to individuals and to the general fabric of our society.

Working alongside government, the educational establishment, public bodies and the business sector, The Suzy Lamplugh Trust has made a significant contribution to personal safety and is increasingly recognised in other parts of the world. The Trust is dedicated to improving personal safety for everyone through research, campaigning, training, consultancy, practical support, educational resources and partnership initiatives.

Verbal abuse and threats constitute a more insidious type of workplace violence than fighting. Bullying can occur between managers and subordinates, colleagues of the same grade, or it can involve the victimisation of one individual by a group. There is no typical situation where bullying takes place.

Increased pressure on staff and managers to meet targets, especially unofficial targets, creates an environment in which intimidation and victimisation are almost unavoidable. While a tough, competitive environment does not create bullies it can certainly aggravate their behaviour. It can also create job insecurity, organisational change and uncertainty, poor working relationships generally and excessive workloads. Unfortunately, harassment and bullying can be seen as strong management, the effective way of getting the job done. Such action by senior managers can be seen as a green light to others to behave in a similar fashion. This can seriously backfire on employers. Staff working in an atmosphere of fear and resentment do not perform well, with the resultant reduction in productivity. Morale levels fall while absenteeism through sickness increases and staff resign. This can result in significant losses in both human and financial resources, costing industry billions of pounds a year in lost labour. It is estimated that millions of working days are lost each year because of bullying at work; more than 160 times the number of days lost in strikes.

The whole question of victimisation at work has been increasingly seen as an important issue throughout Europe. Norway recently improved its Work Environment Act to provide protection for employees. In Finland a study by the Finnish Institute of Occupational Health's department of psychology concluded that psychological terror at work is most conspicuous at workplaces that are characterised by competition and unequal treatment of workers. In Germany, a campaign, Psycho-Terror at the Workplace, has been active for some time and is growing. In Sweden, a new legal provision, which came into force in 1994, outlaws offensive discrimination at work. It made clear that bullying is an organisational issue and that employers have a duty to provide a working environment which will actively discourage bullying.

In this country concern has been growing too. Trade unions have successfully challenged workplace bullies through workplace grievance and disciplinary procedures, industrial action, tribunals and the courts – a move which has focused the minds of employers' liability insurers and employers themselves. But because these existing legal remedies are not specifically designed to deal with harassment and bullying, the position is far from satisfactory.

It appeared vital to The Suzy Lamplugh Trust that, to ensure employers were quite convinced that bullying in the workplace was worthy of their attention, new research was essential to prove that this syndrome was not only harmful to their employees but also that work was likely to suffer. Consequently, The Trust, at the instigation and with the support of Professor Cary Cooper, actively sought sponsorship for a research initiative on 'Destructive Interpersonal Conflict in the Workplace and the role of Management'. This was conducted by Helge Hoel with Charlotte Rayner under the supervision of Professor Cooper. This was most revealing and, as we hoped, both influential and effective.

People who are constantly bullied or harassed lose their self-confidence. Their self-esteem is lowered and their health damaged. Workplace bullying can lead to sleeplessness, migraine, back pain, panic attacks and stress-related illnesses such as depression and anxiety. Nearly three-quarters (72 per cent) of victims of bullying, according to the IPD survey, say that they have suffered from work-related stress in the last five years. This was borne out by the findings of Professor Cary Cooper of UMIST, a leading expert on stress, and also by the Health and Safety Executive in its published guidance for employers on preventing stress at work. This makes clear that bullying can be a cause of the problem and that preventive measures must include action to eliminate bullying wherever it exists.

This book is the end result of the superb work undertaken by Professor Cary Cooper, Helge Hoel of Manchester School of Management, UMIST, and Reader Charlotte Rayner. It is full of accurate information which has not only been analysed but also provides strategies and ideas to help both employers and employees both amicably and productively.

In my view, no public, private or corporate sector can do without this book and every Human Resources Officer should read it and inwardly digest. You will find that neglecting bullying within the workplace can be a very expensive neglect.

Diana Lamplugh OBE
Director, The Suzy Lamplugh Trust
May 2001

1 Introduction

This book sets out to examine the phenomenon of workplace bullying. Although people in workplaces have been dealing with bullying for many years, the concept was named little more than a decade ago. The last ten years have seen considerable professional and academic work on the topic around the world. The purpose of this book is to review current thinking on this difficult topic and in that process we will draw from major writing and research in the field. First we will gear our efforts towards an examination of what we understand about the problem and then we will look toward finding interventions.

Organisational professionals from personnel through to occupational health and trade union staff should find this book of use, and we hope that managers and other interested individuals will find the text informative and useful. Additionally, since it does utilise current research to inform the arguments presented in the text, we expect that academics and students of behaviour at work will use the book as a helpful resource.

The purpose of this book is to inform. As the authors come from academia, we seek information, data, and evidence as our starting point. In this emerging area of bullying at work our 'information' is still far from perfect, and often we will need to show the flaws in our data collection methods and the possible effect on the data and its interpretation. Whilst this book is not a full academic review of the literature, it could be used by those who seek a starting point to take their reading further.

In this Introduction we will try to place our work in context. Why has the issue of bullying at work, which has probably existed for as long as work itself, suddenly become so prominent at this time? Is it an indication that our new millennium represents a fundamental turning point in our on-going industrial revolution?

Or are we dealing with the latest fad which will just go away if we ignore it? Our first section in this chapter will therefore review the development of the topic.

Naturally we need to examine current definitions of bullying, and the basic concepts will be dealt with up-front in this chapter. In addition, we have set aside a later chapter to re-examine the basic concepts introduced here. We will also sketch out the content of the book in this Introduction so that readers can skip sections if they so choose.

Waves of awareness around the world

Across the globe a pattern is emerging of how interest in bullying at work grows. Typically it starts with press reports on some rather dreadful incident of people being treated badly at work. This is followed by a study which exposes the extent to which ordinary people experience negative events in their working lives in that country. The subsequent publicity generates further interest, and often replica or improved studies are undertaken which establish with more certainty that there is 'something' going on (or not) in our working environments.

In Britain the British Broadcasting Corporation (BBC) can take credit for being a lynchpin in our emergent awareness. The BBC ran a radio documentary programme in 1990. In this programme Andrea Adams (a freelance journalist) introduced the topic and played taped interviews of people talking about their negative experiences at work. Despite the fact that the programme was broadcast during the evening which is not a peak listening time, it provoked a strong response from the public. Andrea Adams continued her work on the topic. She wrote an excellent book for targets of bullying which was published in 1992.[1] The BBC continued covering the topic and in 1994 asked her to consult on the production of a TV documentary programme. At this point one of our authors, Charlotte Rayner from Staffordshire University, was brought in to conduct a survey which would provide more formal evidence of the problem. Rayner's subsequent survey found that

---

[1]   We have termed people who get bullied 'targets', but readers should not assume they are necessarily selected in some way – many 'targets' are just in the wrong place at the wrong time and anyone in their position might have their experience.

50 per cent of people thought they had been bullied at some point in their working lives (Rayner 1997). This was a surprise to everyone involved and the study achieved significant publicity.

A similar pattern was reflected in Sweden a decade earlier where the late Heinz Leymann was the key person writing, campaigning and researching the issue. A few years later such processes were witnessed in Norway and Finland. Ståle Einarsen, currently the leading Norwegian researcher in workplace bullying, observed what he called 'waves' of interest where single high-profile incidents (such as surveys or court cases) produced strong surges in attention. Rather like a rising sea tide, the level of interest between the waves has crept to a higher level each time. In Britain, Australia and Germany these waves have produced a continuing and growing tide of interest over the last decade.

The Scandinavian countries are wellknown for their emphasis on egalitarian and caring social values. Therefore it was not surprising that awareness of bullying at school began in these countries and then spread to other countries such as Australia, Britain, Canada, Germany and Japan. A similar geographic pattern has been repeated for workplace bullying. In the USA the study of bullying at work is fragmented, although there is concern about negative behaviour at work and the issue of fairness has received considerable attention.

Early times for support on bullying at work

Whilst the media has played a highly significant role in heightening awareness, a less positive role can be assigned to the management training community and its allied organisations. Mainstream management texts rarely contain any reference to bullying at work, and certainly are scant of suggestions as to what to do about it. In most countries there is no law against bullying, although apparently legal claims under race and gender legislation can have bullying at their base. Research shows that UK laws at least are being stretched to embrace bullying (Earnshaw and Cooper 1996), but this strategy is particularly difficult for white men to pursue as they fall outside the discrimination laws. International law may change this as trading blocks incorporate a social dimension which includes employee rights. As such, the issue typically remains confined to our workplaces and the media, whilst government and the legal profession have yet to deal fully with it.

Why now?

Is the recent development of interest in bullying at work just chance, or are there underlying reasons for growing interest in this topic?

Personal stresses

The last decade has a seen a sharp increase in working hours in some countries such as the UK and USA (Bosch 1999). In these countries this increase has been especially marked amongst managers and professionals who now take work home at weekends as a rule rather than an exception (see Box 1.1).

Box 1.1

- British workers work on average 44.7 hours a week, the highest in Europe.

Source: Rubery, Fagan and Smith (1995).

- 84% of managers believe that their working hours adversely affect their relationship with their children.

- 79% of managers believe that their working hours adversely affect their relationship with their spouse/partner.

Source: Worrall and Cooper (1999).

The single issue of working hours has had a major impact as people struggle to achieve a healthy divide between work and personal lives. Can we have real involvement in a 'life' outside work if we have neither the time nor the energy to engage properly? If our employers are demanding so much from us as we enter the new millennium, do we feel more justified and more able to expect decent treatment when we are at work?

Some areas of work are changing for everyone (see Box 1.2). Today's leaner organisations offer fewer promotion opportunities as layers of management have disappeared. Constant restructuring has removed a sense of job security for many of those who work in management positions in particular (McCarthy *et al.* 1995).

Box 1.2

- The percentage of respondents reporting the use of cost reduction programmes has increased: in PLCs from 63 to 73%; in the public sector from 64 to 69%.
- The percentage of respondents reporting the use of culture change programmes has increased: in PLCs from 54 to 61%; in the public sector from 50 to 57%.

Source: Worrall and Cooper (1999). Reported changes from 1997 to 1999 (Institute of Management).

Certainly we are having to think about long-term careers differently, and this adds to our personal pressure. Such changes in our future prospects can affect us every day. If there are fewer opportunities to advance or a greater chance of being asked to leave, then we all feel a pressure to demonstrate daily our skills and expertise so that we are well placed when selection for promotion or survive the next restructure.

Most people in work have employment contracts which are protected in law. Contracts mean that employees and employers have responsibilities, and both parties also have rights. Running alongside the legal contract, however, is the 'psychological contract' which is an unwritten set of expectations to which an employer and employee adhere (Mullins 1998). The nature of the psychological contract – how much people will exchange in return for employment, and how much employers are willing to provide – is changing. As we have seen, long-term tenure is often no longer on offer from employers (see Box 1.3). Many people are experiencing other shifts by their employer. For example, the responsibility for training and professional updating now often now rests with the employee who has to resource training in their own time or pay their own fees.

Meanwhile at work, those of us in 'industrialised' countries are being put under increasing pressure to justify our existence by adding value to our organisations and their products or services. According to traditional business teaching, 'value' can be added in two ways – either we do things cheaper than our competitors or we do things better. Both need an increase in pace: we (sometimes literally) have to run faster. Not only do we have to do more in the same time, but the quality of our work needs to be good all of the time. The pressure really is on for many individuals.

Box 1.3

- The use of outsourcing has increased: in PLCs from 25 to 31%; in the public sector it has almost doubled from 17 to 30%.
- The use of contract staff has increased: in PLCs from 34 to 39%; in the public sector from 31 to 37%.

Source: Worrall and Cooper (1999). Reported changes from 1997 to 1999 (Institute of Management).

Global scenarios

Whether we work for a large multinational or a small commercial business, competition is often becoming global whether we like it or not. The latest revolution in our industrial development – the convergence of telecommunications – is affecting us all in the shape of e-commerce. Internet connections provide all businesses with opportunities, and most businesses with threats as competitors become more accessible to customers. For many commercial businesses, the pressure to compete on cost is likely to accelerate. As the competitive base widens (perhaps geographically through e-business and e-commerce; perhaps technologically through faster transfer of knowledge, or simply through major corporations seeking new business sectors to buy their way into) so the competitive pressures build. There is a need to find faster and smarter ways of working in order to reduce costs and this means more quality work demanded from all of us more of the time.

Globalisation affects the public sector too as it embraces the concepts of market philosophies. As commercial firms compare themselves (benchmark) with each other, so these competitive ideologies have spread to the public sector. In Britain we have seen a vast increase in benchmarking so that local councils, police forces and education can be transparently measured against their peers. Only a few can ever be at the top of their league. The pressure on 'the rest' will increase, and this will be shouldered by the staff working in those organisations.

Competitive advantage

How do organisations squeeze more out of their resources? Public sector organisations need to be seen to perform well on their key indicators; private sector companies need to out-produce their competitors either on cost or by adding innovative features. In the end, however, organisations don't do things, people do. The last decade has seen a stronger realisation of the importance of people (Senge *et al.* 1999). The notion of 'knowledge' workers has emerged as well as 'knowledge' industries (Handy 1995). Employers have come to understand that the 'learning curve' is critical to success, and so now this no longer forms a source of competitive advantage. Competition has passed to innovations in how we work. Whether innovations are based in science, technology or knowledge, the implementation of those innovations requires teams of people to work together effectively. Fostering innovation and creative attitudes is not enough – we have to implement ideas too!

A confluence of tensions

All this means that in order to do well or just to survive, organisations need capable and motivated people working for them at all levels. These people need to be managed well so that they are pulling in the same direction and are able to handle the complexity of new initiatives. People have always been critical of unfair practices and treatment (e.g. Fine 1985). What is perhaps new is that we are now more critical and unwilling to accept unfair practices, and this is reflected in a growth of litigation around employment issues. But there are other consequences of which we are only just becoming aware.

In this text we will show that when people are treated badly at work they become distanced and many leave their organisation. While employers are likely to push their staff as hard as they can, the 'psychological contract' is two-way in nature. Media coverage perhaps provides the fuel for individuals to label their discontent, and subsequent discussion amongst workers at all levels can raise awareness of the problem (Lewis 1999). At some point people weigh up the costs and the benefits of their working situation, and the costs to each of us can be great (see Box 1.4). Perhaps we are witnessing something new happening at this turn of the millennium? Whether or not people find themselves treated decently is likely to play a role in their evaluation of their working situation.

Box 1.4

- A survey of the European Union's member states found that 28% of employees reported stress-related illness or health problems. This accounts for 41 million EU workers.

Source: European Foundation for the Improvement of Living and Working Conditions (1996).

- A survey by a long-term disability insurer found claims for compensation arising from mental health problems had increased by 90% in the last five years.

Source: Institute of Personnel and Development (1998).

Finally, many Western countries are facing a demographic time-bomb as their populations live longer and the proportion of non-working people increases. Fewer people working may mean employers have to start to compete strongly for those who are available for work. It follows that the need to recruit and retain staff will increase.

What is bullying at work?

Consider the following scenario. A man walks into the office of his personnel department, nervously greets the receptionist and explains quietly that he needs to talk to someone because he thinks he might be being bullied by his boss. He sits down to wait to be seen, often glancing at the door, trying to avoid being seen by those walking past the office.

The next few minutes are crucial for him and the personnel officer by whom he will be interviewed, as some estimate will be made as to whether or not he is being bullied. How will each of them decide? What is it to be 'bullied at work'?

By now there are numerous generalised definitions of bullying at work in organisation policies or handbooks. The following two represent examples of good practice:

> Unwanted behaviour whether physical or verbal which is offensive, humiliating and viewed as unacceptable to the recipient.
>
> (Marks and Spencer)

The misuse of power to intimidate somebody in a way which leaves them feeling hurt, angry, vulnerable or powerless.

(Sheffield City Council)

These generalised definitions can be seen to share some character-istics (IDS 1999). Being bullied at work concerns the experience of treatment by someone else, and therefore it is about behaviours.

Bullying behaviours

There is no definitive list of bullying behaviours. They can include highly observable incidents such as ranting and raving at a person in public. Such behaviours are fairly straightforward to identify and also to corroborate. Similarly, if a group of workers picks on the same person every break by teasing them well beyond a reasonable point, that person might feel bullied. However, bullying behaviours can range towards a much more subtle pattern where, for example, someone alters crucial information which effectively undermines another person doing his or her job. At other times bullying is not so much about what someone does, but about what they don't do. Examples of this could be excluding people by not talking to them, not giving someone crucial information, or not supporting well-earned promotions (see Box 1.5).

Box 1.5

From the moment I dared to make a critical comment, I have been subjected to a campaign of bullying. Not only has she talked about me behind my back and repeatedly humiliated me in public. She has also ensured that no way I will get a future promotion by telling them that I am unstable and unreliable.

Sandra, NHS administrator.

For many examples of bullying, the pattern of behaviours is the key. Taken individually, incidents may seem innocuous, but put together they add up to a scenario which is destabilising and threatening to the person who receives them.

It can be frustrating to those who seek to ensure that their behaviour cannot be construed as bullying that there is no definitive list of negative acts. This situation is unlikely to change. In addition people's reactions to these behaviours vary, and it is to this aspect that we now turn our attention.

Reaction to behaviours

In common with racial and sexual harassment, definitions of workplace bullying extend to the reaction of the person receiving the behaviours. Effectively this takes matters outside the control of the person who does not want to be 'accused' of bullying. Most definitions would include the targets of the behaviour reacting negatively to the behaviour they have experienced. There are a variety of factors around this issue; the nature of the negative reaction, the degree of reaction, and the severity of the effects of the reaction (for example feeling highly irritated or very frightened). While this is a very wide spectrum of potential response, from being very upset by the actions of others through to a full breakdown, it is all negative in nature and effect. Figure 1.1 shows this spectrum.

Figure 1.1 works from left to right and shows how, with repeated negative acts over a period of time, someone might be seen to deteriorate psychologically. What may begin as an ordinary pressure becomes psychological harassment and unabated, repeated

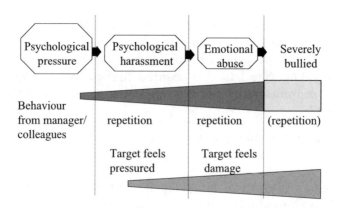

Figure 1.1 A schema of bullying at work.
Source: Rayner 2000a.

negative acts might push the person into feeling abused. In severe trauma people may not need to experience the continuing direct stimulus of negative acts as they mull over the events and effectively repeat their abuse in their own imagination.

Persistency

Early definitions of bullying in the workplace have also stressed persistency as a defining feature. As a concept, persistency is made up of two elements – frequency (i.e. more than one occurrence) and duration (the period of time over which the repeated events happen). These two features have been essential to setting bullying apart from other phenomena such as workplace violence. Persistency was an important feature for the Scandinavian researchers, and their ideas have been the building block for others. For the time being, we would like the reader to accept that most bullying includes repeated behaviours, and we will then return to this point in Chapter 7.

What is to be considered 'frequent'? Is once a week 'frequent' or are we looking for daily events? There are no rules for a practitioner to use. The same problem occurs when we examine the length of the time period (duration) over which one would judge negative experience in order for it to fit together into a 'pattern'. If a group subjected their work-mate to derision every Monday over six weeks, would that be the same as someone whose boss had lost their temper six times in three years? Currently, in order to apply the bullying label, most practitioners would expect some report of repeated behaviours over a period of time, rather than just a single event. As we are becoming more aware of the phenomenon that is bullying, our concepts are changing and this will be explored further in Chapter 7.

Imbalance of power

A final point, included in the Scandinavian definitions but rarely in British ones, is that the target of bullying must be seen to have difficulty in defending themselves. This notion of 'imbalance of power' has had little exposure in Britain, perhaps because it is accepted already. If people have the same degree of power they will be in an ordinary conflict – where two people work out disagreements.

Box 1.6

After repeatedly being attacked in public and undermined, I had had enough and decided to go directly to the bully with my complaint. Unfortunately this got me nowhere. As a result I had to resort to a formal complaint. The formal investigation concluded that the behaviour in question was bullying. Unfortunately, despite a request from myself to have the bully moved to another production unit this was refused by management. Instead I had to continue work next to the person who has made my life a misery. Management response was to hold back the perpetrator's latest salary increment for six months and offer him personal guidance and training.

Karl, factory line worker.

It must be remembered, however, that power can come from many sources. The simple division of formal power and informal power is useful here. Formal power comes from status and position, whilst informal power may be derived from many sources such as the ability to influence others, strength of character, quickness of tongue, etc. Suppose two line managers of the same status are in conflict, but one is very articulate and fast in verbal reactions (therefore strong informal power). Even though both might have the same formal power, their different informal power strengths lead to the existence of different abilities with which to defend themselves.

We shall see that it is difficult to defend oneself against formal power (for example the power which a boss has over a subordinate) as often those in superior positions will support the person with the higher formal status. Their interpretation of events may be believed more readily, and this can be a key issue in cases of bullying where people often need to interpret situations.

Finally, some readers might expect the issue of intent on the side of the 'bully' to be included in a definition. There is a practical difficulty in using this as part of a definition. Expressed simply, if someone accused of bullying states that they did not intend it, then bullying would not have happened if intent were part of the definition. It is quite possible that intent is not present when some people behave negatively to others. Their focus may be exclusively on getting a job achieved (rather than considering how it is achieved) or they maybe simply unaware that their behaviour is experienced as bullying. These are important considerations for those who deal with a case of bullying. Intent, however, cannot be part of the definition as then the whole definition would rest on it. In summary we would suggest that intent is an issue important to a target and essential when considering the interventions necessary, but cannot be included in a definition.

A straightforward definition of bullying at work

The definitions used in most organisations provide an agenda for the personnel officer. The person who is going to meet the client we introduced earlier will first be trying to establish what is going on. They will expect that the person who says he has been bullied can report negative behaviours to which he has been subjected. There may be some behaviour such as a public 'dressing down' which is unambiguous, but the behaviour may not be so obvious. Instead he or she may present the personnel officer with a string of rather minor incidents but when a full picture emerges, a consistent pattern is found. In addition our complainant will be having some sort of negative reaction to the experience of the behaviours.

The Scandinavian approach

Scandinavian researchers began studying bullying at work relatively early and have established a sophisticated approach which is worth considering. Instead of 'adding up' the parameters (such as frequency, duration and reaction) they have a definition of bullying based on a process of escalating conflict.

For example, a man in an accounts department failed to get the promotion he expected and thought he deserved. He saw the manager (who had not recommended the promotion) as focusing on only a small part of his job and not taking into account all the extra work he had put into other areas. The manager explained that the promotion was connected only with certain facets of the job, and at a much higher level than currently performed by the employee. After several unsuccessful attempts to appeal against the decision, the employee decided to withdraw the extra energy he spent on the job. In effect he just did as he was told. The manager found that jobs were not completed to the previous standard and became frustrated. It became a daily event for the manager to confront the man on the standard of his work, to which our employee would reply that he had just done what he had been told. Some months later the employee was asked to attend a disciplinary interview where he received a warning on the professionalism of his work. Then the manager downgraded the work that the employee was given (in the manager's eyes) to limit any damage caused by this member of staff. The employee went to his trade union representative to make a formal complaint about his treatment and the low level of work he had to do. At the same time he took evidence from his doctor to show that he was suffering from stress at work. This sequence of events is a small example of an event causing repercussions well beyond that which one would envisage. In it, we can see an example of how conflict escalates. Trade union representatives in the UK often report that interpersonal conflicts have escalated out of proportion and gone beyond reconciliation by the time they are made aware of them.

A case for multiple definitions?

Simply, bullying may be defined differently by different people and professions. Consider the lawyer who is trying to decide whether they have a case to bring. They will be looking for repeated behaviours, ideally with evidence of their occurrence corroborated by independent others. They will also be looking for substantial damage to the individual which will form the reason for a case.

These would not be the criteria that one would hope a personnel officer would use to 'trip' their alert mechanisms. Personnel staff, managers, trade union representatives and occupational health professionals are at a different place in the 'alert' chain and they should have a lower threshold, otherwise they will always be ending up in court! These staff will have to choose the point at which they will become concerned and also the point at which they will decide that investigation or intervention is necessary. Will they become concerned when they hear about any bullying behaviours?

Currently, most personnel officers will act on a report of sexual or racial harassment whether or not employees have reacted to it because they know that the next time it happens it *might* cause distress to someone. The base of their caring may only be in keeping the employer out of courts – but they will react. 'Duty of care' is a fundamental principle that all employers have to observe, as are other aspects of health and safety legislation. Since it has been established that bullying at work can damage employees, many personnel professionals treat bullying as they would treat health and safety hazards – the problem being that people are not as easy to deal with as faulty fire extinguishers!

Box 1.7

When I questioned a previous departmental decision, he made it clear that I was free to leave at any time. However, when I stated that was not my intention, he looked at me with contempt, telling me that I might come to regret that decision later.

Jeanette, bank employee.

Professionals within the organisation will also need to examine the 'organisation-wide' level of alert. Individual actions and single events combine to provide a broader picture of what is going on in their place of work. Assessing the full picture may produce concern at a level beyond that of the individuals reporting difficulties. Of course professionals will be concerned about being taken to court, but from a business point of view they will also have a watching brief to ensure effectiveness and efficiency which will be threatened by a climate of bullying.

Targets of bullying may also struggle with a definition. They may not be aware of the label, or they may think that the label applies only to the school playground. In other cases they may feel ashamed about being bullied and resist applying the label to themselves. It can be unhelpful for targets of bullying to have such vague operational definitions with which to identify their experience. Perhaps the most important thing for targets is their reaction – they will be concerned when they are having a negative reaction. They may be concerned only about behaviour to which they personally react, and may become highly sensitive to the point of being worried about being judged as paranoid.

Those who wish to ensure that their behaviour is not bullying do not have a definitive list of behaviours which they need to avoid. In addition, their own judgement of 'what is bullying' has only limited relevance as it is the reaction of the recipient, which is crucial in bullying incidents.

At the organisational level, a different agenda may be in operation which relates to the competitive issues discussed earlier, and which may involve the employing organisation simply surviving. The need to retain good staff, and the need to provide an environment within which staff can contribute openly to the organisation is essential. Strategic decision makers may become aware of climates of fear and bullying and know their negative effect on openness and the quality of information they are given to work with. Eventually the decision makers may also become aware of the huge amount of resources that go into dealing with bullying at work – resources which could be put to much more positive use.

What comes next?

We have tried to make our text logical in its progression as we explore the phenomenon of workplace bullying. The next chapter looks at data relating to *who* gets bullied. It is followed by a chapter which identifies the *consequences* of bullying. These chapters draw heavily on a recent study by two of our authors, Helge Hoel and Cary Cooper from the University of Manchester Institute of Science and Technology (UMIST). This study is currently the most comprehensive undertaken in Britain. Our focus then shifts from the targets of bullying to the *bullies* (Chapter 4) and to the *organisation* (Chapter 5) where we will examine the role of corporate culture. These issues are brought together in Chapter 6, which examines the interactions between the individual players and the organisation in order to determine the *instigators* of bullying at work.

By Chapter 7 the reader will have a greater appreciation of the subtleties presented by bullying at work, and this short chapter revisits some of our concepts of bullying at work and examines issues that both perplex and remain to be tackled.

Later chapters turn toward making changes. Chapter 8 examines how blame for bullying is treated in the organisation. As such it introduces our discussion on how to tackle bullying. In Chapter 9 we concentrate on what *individuals* can do, while Chapter 10 places the focus for intervention at the *organisational* level. We finish our text by turning toward the future and outline an agenda for further work and change.

2 The targets of bullying

In order to stimulate action to deal with bullying at work, it is essential to have a clear idea of the scale of the problem. This chapter reports the most up-to-date findings on the prevalence of bullying in Britain. It will become evident that bullying is a very significant workplace problem affecting a large number of people. This is particularly the case when we widen our perspective to include those who are affected indirectly, such as the bystanders of bullying incidents. Particular industries and occupations that seem to be 'high-risk areas' for bullying will be identified in this chapter. We will also examine how factors such as gender, age, and ethnicity might impinge upon the experience of bullying. Do people who are of a certain age, for example, get bullied more, or less? Do managers get bullied more than front-line workers? For how long are people bullied? Do people get bullied on their own, or in groups?

Bullying is not an abstract phenomenon but relates to the experience of negative behaviours. We will take a closer look at these behaviours, explore what they may have in common and how frequently people experience them. Individuals react differently, and we will see how they try to deal with their experience.

Studying workplace bullying

This chapter will draw extensively on studies of bullying undertaken by the authors. In a BBC-sponsored study of workplace bullying, the first of its kind in the UK, Charlotte Rayner surveyed over 1,100 part-time students at Staffordshire University (Rayner 1997). She followed this with two studies in co-operation with the public employee union UNISON (1997, 2000). Finally, a recent study undertaken by Helge Hoel and Cary Cooper at UMIST involved more than 5,000 employees from over 70 organisations in the public, private and the voluntary sectors across Great Britain (Hoel and Cooper 2000). As this study is the most comprehensive to date, it will be examined in detail so that readers can get a full picture of how such studies are undertaken and where the data comes from.

The first problem for researchers is finding people to participate. The UMIST study was the first in Britain to involve a group of employers (rather than, for example, trade unions). Central to the success of the project was an organisation called BOHRF (British Occupational Health Research Foundation) which comprises a group of employers (sponsors) together with representatives from the Trade Union Congress (TUC), the Confederation of British Industry (CBI) and the Health and Safety Executive (HSE). BOHRF fund research projects on occupational health issue, such as the UMIST study.

Several BOHRF member organisations participated, and their initial interest encouraged other organisations to join the study. Five trade unions or staff associations took part: BIFU and UNIFY (banking sector), NASUWT (teachers), CWU (post and tele-communications) and the Police Federation (police officers). In total 70 organisations participated which, together, employ close to one million people. It constitutes the primary source of information for this chapter.

A geographic restriction was placed on the sample for the study. Only employees who worked in Great Britain (i.e. England, Scotland and Wales) were included. The political situation in Northern Ireland and the possible influence of what may be termed 'sectarian harassment' limited the study to Great Britain (Hoel and Cooper 2000).

In order to make generalisations from the UMIST study findings, a representative sample of staff from participating organisations was required. To meet this objective, a procedure was developed, an example of which is given in Appendix 1.

During the spring of 1999 a total of 12,350 questionnaire packs were distributed throughout the participating organisations. Each 'pack' contained a questionnaire and an envelope so that replies could be sent straight back to the researchers. This procedure helped to meet the guarantees of anonymity and confidentiality which were emphasised in order to encourage people to respond.

More than 5,300 questionnaires were returned to the researchers, of which 5,288 were used for analysis (the others being 'spoiled' or blank). Thus there was a usable response-rate of 43 per cent. Taking

Table 2.1 Sample response rates

Sector	Number sent out	Number returned	Response rate (%)
NHS trusts	1,069	535	51.0
Post & telecommunications	1,000	273	27.0
Civil service	250	141	56.0
Higher education	1,072	487	45.0
Teaching	1,000	426	43.0
Local authority	924	388	42.0
Manufacturing industry	1,162	536	46.0
Hotels	493	163	33.0
Retailing	855	354	41.0
Banking	820	262	32.0
Voluntary organisations*	317	123	39.0
Dance	196	85	43.0
Police service	1,000	483	48.0
Fire service	1,167	520	45.0
Prison service	1,000	471	47.0
Total sample	12,350	5,288	42.8

Source: Hoel and Cooper 2000 (numbers have been rounded up at 0.5).
Note: *The voluntary sector sample includes housing associations, trade unions and charities.

into account the sensitivity of the issue, the high response-rate strengthened the reliability of the study findings. An overview of the sample response-rates is given in Table 2.1.

The sample for this survey can be seen to represent many occupations in both the public and private sectors. A summary of the industrial groups responding to the sample is shown in Figure 2.1.

Demographics of sample respondents

The age profile of respondents in this study followed a normal distribution curve with the majority of respondents falling within the middle age groups and fewer respondents amongst the younger and older age groups. The average age was 43. An even gender split was achieved with 52.5 per cent of the respondents being men and 47.5 per cent women. A large majority of the respondent sample (97.1 per cent) defined themselves as white. A total of 85 per cent worked full-time as opposed to 15 per cent who worked part-time and the sample reflects UK norms for employment in these respects (Hoel and Cooper 2000).

The sample had been stratified so that a good number of managers and professionals responded. Nearly half the respondents

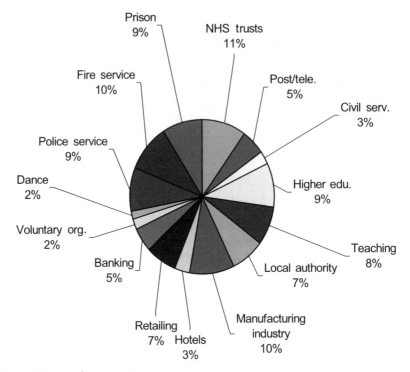

Figure 2.1 Sample composition.
Source: Hoel and Cooper 2000.

(47.5 per cent) defined their job as professional or managerial. Whilst 43.6 per cent saw themselves as workers, 43.3 per cent reported to have managerial responsibility of some kind. This allowed testing to be done on the managerial and professional sample, as there were sufficient respondents. We will now turn our attention to the results of the survey.

Is there really a problem?

In the Introduction we showed how the media have played a significant role in highlighting workplace bullying. Is it possible that the entire bullying problem is a media creation? The fact that the issue didn't appear on the scene in the UK before in the early 1990s could suggest that this is the case. However, a steady stream of studies undertaken in the UK since the mid-1990s have continued to produce results which strongly suggests that the media are not making things up. Table 2.2 shows some of these surveys.

Table 2.2 British surveys of bullying at work

Author	Date	No.	Sample	Incidence
IPD	1996	1,000	Telephone survey	12.5% In last 5 years
Rayner, C.	1997	1,137	Part-time students	53% During working life
UNISON	1997	736	Public sector union members	14% In last 6 months
TUC	1998	1,000+	NOP telephone survey	11% In last 6 months
Lewis, D.	1999	415	Union members further/higher education	18% In last 6 months
Quine, L.	1999	1,100	Employee of NHS community trust	38% In last year

The final column of Table 2.2 illustrates how studies often use different periods of time over which to measure bullying, which makes direct comparison impossible as it affects how bullying is defined for any study. For example, Rayner, in her study of part-time students, asked about the whole of respondents' working lives. Quine's study of NHS employees measured the response from those who had been bullied within the last year. To come into line with other European studies, Rayner in her UNISON surveys (1997, 2000) measured bullying over the previous six months. This strategy was followed by Hoel and Cooper as one of their central objectives was to establish, using a structured sample, the prevalence of the problem in different sectors and occupations of British workplaces.

Asking if people have been bullied

Since people are likely to include different things within the label 'bullying', Hoel and Cooper, like other researchers, used a definition in their study. Most studies have placed the definition in a covering letter or right at the start of the questionnaire. Hoel and Cooper decided to place their definition of bullying inside the questionnaire and right next to the question which asked respondents if they had been bullied. The definition they used was based on a Scandinavian definition (Einarsen and Skogstad 1996) which emphasised the negative, persistent and long-term nature of the experience.

> We define bullying as a situation where one or several individuals persistently over a period of time perceive themselves to be on the receiving end of negative actions from one or several persons, in a situation where the target of bullying has difficulty in defending him or herself against these actions. We will not refer to a one-off incident as bullying.
>
> (Hoel and Cooper 2000)

Many studies of bullying have simply asked the respondents whether they have been bullied at work within a particular timeframe with a simple Yes/No answer. To establish the scale and intensity of the bullying experience, a question immediately followed the definition:

> Using the above definition, please state whether you have been bullied at work over the last six months.

Instead of a simple Yes/No response, several options were given which allowed respondents to show how frequently they felt bullied. Figure 2.2 shows the response categories and the overall results.

A total of 553 respondents (10.6 per cent) reported that they had been bullied over the last six months. For one in seven of those labelling themselves as being bullied, the experience had taken place on a regular basis with at least one negative encounter per week.

This question provided a comparison with other studies that measured bullying over six months. Respondents were also asked if they had been bullied in the last five years. When the time period was extended to the last five years, nearly a quarter of all respondents or 24.7 per cent reported being bullied. This figure also includes those bullied within the last six months. If we include those who have experienced bullying as a witness or a bystander, we may conclude that nearly half of respondents in the study have had some experience of bullying, either directly or indirectly within the last five years, and this confirms previous UK studies.

Given these figures, how serious a work problem is bullying? The current number of employees in the UK is currently estimated at 24 millions (TUC 2000). The figures from Hoel and Cooper's national study provided data that close to 2.5 million people can be

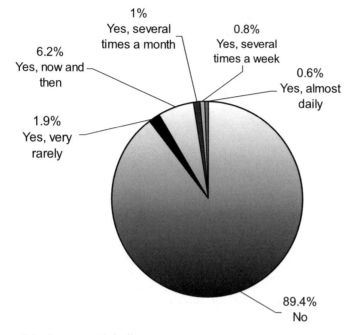

Figure 2.2 Contact with bullying.
Source: Hoel and Cooper 2000.

considered as labelling themselves as having been bullied during the last six months.

If we concentrate on the most severely victimised group, those who have been bullied on a weekly or daily basis (1.4 per cent of the total sample), we are possibly talking about close to 300,000 individuals. The Hoel and Cooper figures provide a background against which one can conclude that bullying is a significant work-problem of a disturbing magnitude. If we extend the time-frame of experience we may conclude that bullying is likely to affect the lives of a majority of us either directly or indirectly.

Identifying high-risk groups

One of the objectives of the UMIST study was to compare the incidence of bullying between sectors. It will come as no surprise to some readers that bullying is more evident in some sectors and occupations than in others. An overview of prevalence figures for everyone who considered themselves bullied across occupational sectors can be found in Table 2.3.

Table 2.3 Prevalence of bullying by sector

Sector	Bullied in last 6 months (%)	Bullied in last 5 years (%)	Witnessed bullying in last 5 years (%)
Post/telecommunications	16	28	50
Prison service	16	32	64
Teaching	16	36	58
Other	14	20	40
Dance	14	30	50
Police service	12	30	46
Banking	12	25	40
Voluntary organisations	11	27	56
NHS trusts	11	25	47
Local authority	10	21	43
Civil service	10	26	47
Fire service	9	20	43
Hotel industry	7	17	46
Higher education	7	21	43
Retailing	7	18	34
Manufacturing	4	19	39
Totals	10.6	24.7	46.5

Source: Hoel and Cooper 2000.

From Table 2.3 we should be able to draw some conclusions. Of the sectors surveyed, the high-risk sectors or occupations appear to be: the prison service, post and telecommunications, teachers and the dance profession. At the other end of the spectrum we find manufacturing. However, before reading too much into these differences we would like to emphasise that most of the samples at the lower end of the prevalence spectrum were less representative than some of the samples from the top end of the spectrum. For example, the brewing and pharmaceutical companies in manufacturing were represented predominantly by sales and marketing staff. The prison service was better represented with a random sample taken from the entire service from top to bottom of the organisational pyramid.

We should also notice that whilst figures for reported bullying in some sectors may be considered low, the number of people previously bullied and those having witnessed bullying may indicate that the problem may be greater than first anticipated. Take the manufacturing industry as an example. Whilst only 4.1 per cent of respondents reported being bullied the last six months, 19.2 per cent reported being bullied within the last five years, raising to 39.0 per cent for those who stated that they had witnessed bullying. So whilst relatively few reported a recent experience of being bullied themselves, four out of ten of all staff had observed it taking place within the last five years.

How can we explain the figures in the high-risk sectors? In the case of the prison service, the presence of internal organisational problems and conflicts, particularly between management and staff and between different groups of employees have been highlighted (e.g. Power *et al.* 1997). Prison staff have been accused of using unacceptable methods, possibly of a bullying nature, against inmates or other members of staff (*Guardian* 2000). In addition, the 'bullying label' has been used actively by prison staff for a number of years in order to describe relationships between inmates. These factors may make it easier for staff to use the label, and to consider negative behaviour to be bullying.

Explaining specific differences may be a topic for other studies and here we would be concerned that placing too strong a focus on the 'top league' of bullying may deflect from the more important finding that bullying appears to be a problem throughout the workplace.

Bullying as a drawn-out affair

Various studies have investigated the duration of bullying. In the UMIST study the researchers focused in particular on the long-term duration of exposure to negative behaviour and its association with the damage that bullying seems to inflict on targets (see Chapter 3). Figure 2.3 shows data from the UMIST study. Readers' attention is drawn to the high number of people for whom the bullying has been going on longer than two years.

For some respondents the episode that began the bullying process may be easy to remember. In such cases it may be relatively straightforward to estimate how long the bullying has gone on. However, other people may find it difficult to identify exactly when the bullying started, so we cannot be certain of the validity of these reports.

In addition, routine behaviour and behaviour which at the time was considered relatively innocent or unintended, may over time (with repetition and in different circumstances) take on a different meaning for the recipient. Despite the problems in establishing the validity of the data in Figure 2.3, the figures tell us that bullying is a prolonged problem which may persevere for a very long time. In the

Hoel and Cooper 2000

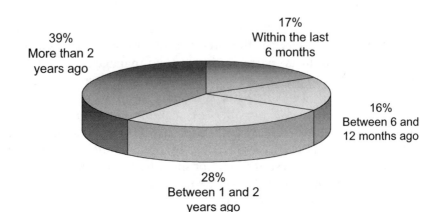

Figure 2.3 Duration of bullying.

authors' experience it is not that unusual to come across cases which have continued for five or even ten years.

Bullying and gender differences

Gender differences were investigated for incidence. When looking at the results from people who said they had been bullied in the last six months, no statistically significant differences were found between men and women (at the 0.05 per cent level).

This confirms previous research into bullying which has consistently found that women and men are bullied at about the same rate.

However, when experience over a five-year time period was considered, women were found to be statistically more likely than men to report being bullied (p<0.001) in the Hoel and Cooper study where 28 per cent of women and 22 per cent of the men reported being bullied 'In the last five years'. How can we explain the difference found after five years, considering there was no statistically significant difference in the gender of those who are reporting being bullied at the time of the research? In general women reported a higher frequency of experience of behaviours than men, with 1.7 per cent of women as opposed to 1.1 per cent for men reporting exposure to bullying behaviours on a weekly or daily basis. Therefore it is possible that the issue is one of labelling. However, why women do not label their treatment as being 'bullied' close to the time when it is going on, but do label it later is not known. A possible explanation for this difference may be that women remember incidents better or perhaps that women reflect more on past experience. We will return to the issue of gender and bullying when we examine what behaviours people experience when they are bullied.

Bullying and age

Age seems to be of relatively minor importance when it comes to bullying. In the UMIST study, younger employees reported the highest level of bullying, followed by those in the 35–44 age band. Those in the age bracket above 55 years appeared to be least likely to report being bullied. This result reflects the findings from previous UK studies, e.g. Rayner (1997).

In some countries (e.g. Norway) the opposite has been found, where older employees report the highest incidence. This has been explained by the local researchers using anecdotal data and included the possibilities that older people may work at a slower pace and sometimes may be reluctant to change (Einarsen *et al.* 1994a). The situation is complex and individual differences will vary, as will circumstance. One could argue that that older employees may have a clear opinion that they deserve to be treated with respect and that this may also render them vulnerable or sensitive to bullying. On the other hand, work experience, knowledge of the organisation and personal networks may all work against bullying and victimisation, making older employees less likely targets. It is reasonable to assume that different dynamics are at work within different national cultures, and this may be worthy of further investigation.

The relatively high prevalence of bullying among the youngest employees in Britain may be explained by referring to the relatively low age at which young people enter employment in the UK. This factor may increase an individual's vulnerability to bullying (Hoel *et al.* 1999). In the same way, low self-esteem and personal insecurity may increase an individual's likelihood of becoming targeted (Adams 1992), being over-confident may also at times represent a problem. This is particularly true if the individual finds it hard to grasp the norms of the organisation. Failure to observe such norms may result in isolation and social rejection by the work group (Schuster 1996). However, we need to do more research in order to understand what is going on.

British studies have found that, as a general trend, the experience of being bullied tended to last longer the older you are. So whilst only 9 per cent of 16–24 year olds had been bullied for more than two years, the response for those over 55 years was 51 per cent in the Hoel and Cooper study. The fact that younger employees may change their jobs more frequently than older ones may account for some of this difference. These findings suggest that incidents of bullying become increasingly difficult to resolve the older the target. Reduced opportunity to find alternative work may be a possible explanation, leaving targets with no other option than staying put and coping as best they can.

Bullying and ethnicity

We know little (so far) about the relationship between ethnicity and bullying. No study has looked effectively into this particular issue. Even in the UMIST study, less than 3 per cent of respondents came from ethnic minorities. This represents less than 200 people, so making any generalisation is difficult. Asians were found to be over-represented among targets of bullying with 19.6 per cent of respondents, as opposed to 10.5 per cent of respondents identifying themselves as 'white' reported being bullied. However, as the actual number of respondents was small, we would not want to place significance in these findings except to suggest they point to the need for further research.

Bullying and job tenure

Intuitively one may think holding a part-time job or being employed on a fixed-term contract would make one more vulnerable to becoming a target of bullying. However, the UMIST survey findings do not support this suggestion. According to the survey, data staff on full-time contract were found to report being bullied more than part-time workers (see Figure 2.4). It should be emphasised that the large majority of survey participants held a permanent full-time contract.

Having a looser relationship with the organisation in terms of time spent at work and future relationship with the organisation may function as a kind of buffer against being bullied (Hoel and Cooper 2000). These people may be less well known to the bully or bullies, and knowing less about a potential target's reaction and possible retaliation may prevent a bully from attacking in the first place.

In cases where the perpetrator finds gratification by feeling powerful and in control, the lack of physical availability of people without full-time permanent contracts may make them less attractive. The general vulnerability of part-timers and people on fixed-term contracts may also make such employees more careful in trying to avoid conflict in the first place.

Previous UK studies have drawn their samples from employees with little managerial responsibility. It is not known whether this has been a function of lack of access to managers or the researchers assuming that front-line staff were most at risk from being bullied. The Hoel and Cooper study was the first to investigate managers properly. It was surprising to find that bullying appears to be relatively equally distributed across the organisational hierarchy, including senior managers as shown in Figure 2.5.

Whilst the prevalence of self-reported bullying may be the same across the organisation, the distribution of negative behaviours may not. It is likely that the bullying experience of a supervisor may be of a more direct and overt nature than that of a senior manager. At the higher levels control over rewards in the form of career progression, salary and organisational standing may be important factors.

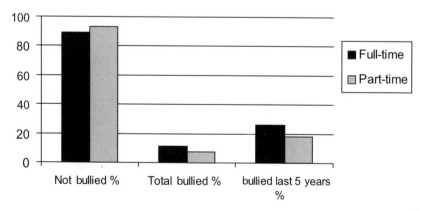

Figure 2.4 Bullying and contracted hours.
Source: Hoel and Cooper 2000.

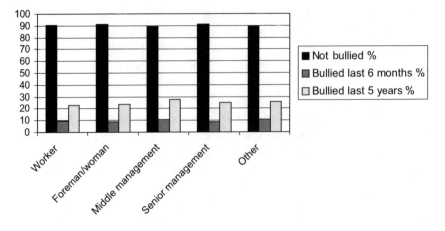

Figure 2.5 Bullying in the hierarchy.
Source: Hoel and Cooper 2000.

Interestingly, the group in the study that most frequently reported to have witnessed bullying without being bullied themselves were middle and senior managers. That these managers admitted to having witnessed bullying is a further indication that the problem is widespread within organisations and cannot be written off as merely an expression of dissatisfaction on the part of disgruntled employees.

Being bullied in groups

Previous studies have found that there is a wide variation in how many people are bullied together; how many people are singled out and bullied on their own, to where the whole of the workgroup is bullied. For example, the first UNISON study (1997) had a response base from the whole trade union (which has many NHS workers). It found few people singled out, and around a third bullied with the whole of their group. The pattern changed in the later survey (UNISON 2000), which was conducted exclusively amongst civilian workers in the British police force. More people were found to be singled out and also more people reported the whole of their group to be bullied. This may reflect the high level of team-working in the police environment. The Hoel and Cooper study found a further pattern where more people reported being singled out than had been shown before (see Figure 2.6). This could again reflect the sample,

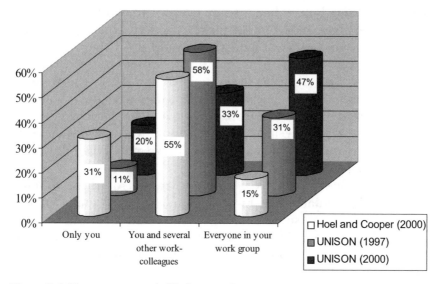

Figure 2.6 How many were bullied at once?

which had a relatively high number of managers who may not see themselves as part of a 'workgroup' in the same way that UNISON members might.

What is interesting about Figure 2.6 is that the majority of people report being bullied with other colleagues. Perhaps where everyone is being exposed to bullying, we might term this a 'regime'. It should be stressed that, although a number of respondents reported a shared experience of being targeted with one other or several colleagues, bullying is still likely to be an isolating experience.

If we focus on particular sectors and occupations, then being singled out for mistreatment was most prevalent in the police, the fire service and the retail sector. As far as the uniformed services were concerned, this phenomenon may be explained by reference to the organisational culture (see Archer 1999) which has strong group bonding and where penalties for breaking ranks are strong.

What do targets do when faced with bullying?

This is an important question to tackle as the results can feed into our prevention strategies on workplace bullying. The UNISON study in 1997 (which had a broad sample of public sector workers taking part) was the first to ask this question and found that those who were being bullied had already tried a variety of routes. In this study targets were asked about their actions and other participants were asked what they thought they would do if they were to be bullied. As can be seen from Table 2.4, those who were actually being bullied had taken far fewer actions than those who were not being bullied imagined they would take if they were to be bullied.

Of course these results are self-reports and we do not know how reliable they are. They indicate that people who are not currently being bullied anticipate much more proactive moves than those who are actually being bullied take.

These studies also looked into the results of these different actions. The actual numbers of respondents were quite small in different subgroups, so we can only draw tentative conclusions. The most common report from respondents to the survey for all actions was that 'nothing' happened as a result. It is quite possible that targets do not hear of actions that people make on their behalf, so we do not know that in reality 'nothing' did happen. It is important for us to grasp that their perception is that nothing happened though, and this has validity in itself.

When something did happen, in several cases (i.e. confronting the bully, going to the bully's boss, or making a group complaint) the situation was more often reported to have got worse than improving. This highlights the need for targets to get support when making confrontations. In general, those targets who did go to professionals such as the trade union or personnel staff reported

Table 2.4 The actions of the bullied and the non-bullied. Comparing the actions of those who were being bullied and the anticipated actions of those who were not currently being bullied

Action	Not currently bullied (%)	Currently bullied (%)
Confront the bully	73	60
Go to bully's boss	63	46
Consult Personnel	55	24
Go to Occupational Health	23	5
See UNISON rep.	73	26
Get support of others to complain	60	21
Leave your job	7	36 (intend)
Stay in job + do nothing	5	31 (intend)

Source: UNISON 1997.

few negative outcomes, and this seems a safer route than acting alone which can have high risks.

The UNISON study in 1997 only asked about the formal actions taken by targets of bullying. The Hoel and Cooper study in 2000 extended the enquiry to include less formal actions such as talking to colleagues and family. The results are shown in Figure 2.7, where it can be seen that such informal actions were frequent. The action most commonly reported by targets of bullying was 'discussing the problem with work colleagues', followed by 'discussing the problem with friends and family'.

Women appeared to be more active in seeking support when being bullied; two-thirds of women talked the case over with family and friends, whilst only a third of men utilised this channel of support. Actions people took did vary, for example 34 per cent of all targets reported that they 'confronted the bully', and a total of 13 per cent responded that they 'did nothing'. This data illustrates that for many individuals neither personnel nor the union /staff association seemed to be viable options when they are faced with bullying – although obviously we are making an assumption here. Occupational health services were used rarely, with medical advice being sought more often from GPs.

The latest UNISON study in 2000 looked at civilian workers in the British police force. This study put an 'open-ended' question to people who had been bullied before but were not currently being bullied. This question requested information from anyone who had successfully stopped being bullied. The replies were varied in con-

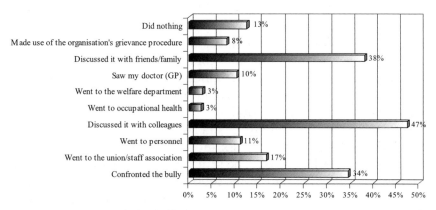

Figure 2.7 Actions on being bullied.
Source: Hoel and Cooper 2000.

tent, but one element came through in the analysis. Targets who had successfully made the bullying stop took their action very quickly – as soon as the negative behaviour occurred. The conclusion drawn from this result was that these people had stopped the bullying become an established pattern by voicing their non-acceptance of it very fast. Of course the 'bully' would have needed to be receptive to the message, but perhaps through giving the message of unacceptability very quickly, the message is clear to hear.

Measuring bullying

Traditionally bullying has been measured in two different ways. Some, including the authors of this book (e.g. Hoel and Cooper 2000; Rayner 1997) have followed the approach suggested by Einarsen and his colleagues in Norway (Einarsen and Skogstad 1996). This method presents questionnaire participants with a definition of bullying and asks whether this description would apply to their experience. In other words, we want to establish how many people perceive and label themselves as being targets of bullying.

A second approach was chosen by Leymann (e.g. Leymann 1996). He presented study participants with a list of negative behaviours. If people reported that they had experienced at least one of these behaviours over a period of six months on a weekly or more frequent basis, he concluded that the person was bullied. He did not ask respondents to label themselves. It could be argued that this is a more objective method of measuring bullying than the more subjective 'labelling' method.

The first UNISON survey in 1997 compared the two methods. Around half the people who said they had experienced frequent bullying behaviours did not label themselves as bullied (Rayner 1999b), as can be seen in the first column of Table 2.5.

Using the first (labelling) definition one would accept everyone in the Bullied? 'Yes' line, and would include any frequency criteria which would provide a total of 135 (103+18+14). Using the second (behaviours) definition one would accept people in cells Bullied? 'Yes' and 'No' over six months and on at least a weekly basis – i.e. 103+111=214. The same exercise was carried out with the second UNISON study in 2000 and found the same results. Each method produces different numbers but also includes different people in the sets. This highlights how careful we must be

Table 2.5 Experience of negative behaviour and its relationship to labelling. The numbers show the actual number of people labelling themselves as being bullied, and the frequency of behaviours they experienced in the UNISON 1997 survey

Experience of behaviours	Daily+ weekly	Every month	Under monthly	Never
Bullied? 'YES'	103	18	14	4
Bullied? 'NO'	111	79	160	248

Source: Rayner 1999b.
Note: N=737 with 24 responses having incomplete or missing data – i.e. 761 responses to questionnaire.

when comparing studies, and has methodology implications for all studies into bullying at work.

What bullying behaviours are reported?

Hoel and Cooper (2000) decided also to use both types of labelling and ask about the frequency of behaviours.

A questionnaire developed in Norway by Einarsen and Raknes (1997), the Negative Acts Questionnaire, was a good starting point. It was used, with the original authors' permission, in group discussions about bullying which were held in different UK workplaces and industrial sectors in order to generate the questionnaire. An overview of the questions that were finally used for negative behaviour can be found in Appendix 2.

Top-ranked negative behaviours

Respondents were asked to identify how frequent they were exposed to any of 29 negative behaviours within the last six months (see Appendix 2). Respondents were given the following answer alternatives: 'never', 'now and then', 'monthly', 'weekly' and 'daily'. Responses for all of the items can be found in Appendix 2.

The negative behaviour most frequently reported by respondents was 'someone withholding information which affects your performance', which confirms previous studies (Einarsen and Raknes 1997). A total of 67 per cent of respondents reported having been exposed to such behaviour. The second most reported behaviour was 'having your opinions and views ignored', which was experienced by 57 per cent of respondents. Two further questionnaire items received an endorsement of above 50 per cent: 'being exposed to an unmanageable workload' (54 per cent) and 'being given tasks with unreasonable or impossible deadlines' (52 per cent). Readers will note that these are all quite indirect actions.

At the opposite end of the frequency spectrum were found direct forms of behaviour. The behaviours least frequently encountered were 'offensive remarks or behaviour with reference to your race or ethnicity' (but there were few respondents from ethnic minorities) followed by 'insulting messages, telephone calls or e-mails'.

If people reported 'weekly' or 'daily' exposure to a behaviour this was termed 'regular exposure', and less frequent exposure was termed 'occasional experience'. Figure 2.8 shows the top categories. Indirect acts (such as information manipulation) are much more commonly found than direct acts (such as being shouted at).

The issue of frequency is important, as bullying is often about the repetition of small acts which individually mean little, but together form a pattern with which individuals find difficulty coping. We all experience the occasional unmanageable workload, but here the respondents are reporting that it is a constant pressure.

People's control over their working situation and the rewards they get from coping with the pressure may also play a role. If the negative situation is seen as being temporary and one is likely to be rewarded by keeping going then someone's perception will be different than if they feel that the situation is unlikely to change in the foreseeable future and it is outside their control.

Physical violence

The study confirms previous findings (e.g. UNISON 1997) which suggest that physical violence in the workplace is still relatively uncommon in Britain, with typically over 99 per cent of people reporting no frequency at all. There were some worrying pockets of responses. In the prison service nearly a third (32 per cent) of respondents to the study reported having received 'threats of violence or physical abuse' in the last six months. In addition, NHS respondents and the teaching profession reported having faced such behaviour.

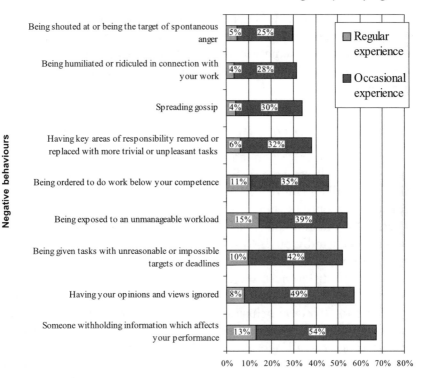

Figure 2.8 Most frequently reported negative behaviours.
Source: Hoel and Cooper 2000.

Sexual harassment

The study incorporated one item relating to sexual harassment – being subjected to 'unwanted sexual attention'. Overall around one in ten had experienced this form of sexual harassment within the last six months, of whom 70 per cent were women. While few (1.2 per cent) reported regular experience, it varied greatly between industrial sectors and occupational groups. For example, a total of 24 per cent of respondents reported this type of harassment in the hotel industry, followed by 15 per cent and 14 per cent respectively for the NHS and the prison service. Only 4 per cent was reported by employees of the banking industry (Hoel and Cooper 2000).

The large discrepancy between these figures may in part be explained by reference to the varying nature of work across sectors where, for example, in the case of the NHS and the hotel industry such incidents might take place in encounters with client groups. Such an argument may not be as convincing for the prison service. In this case the explanation is more likely to be found within the organisation itself where behaviour norms may be far more important.

Variation in patterns of exposure to negative behaviours

Generally, younger respondents reported more negative behaviours, and readers will remember that younger respondents labelled themselves as 'bullied' more often than older workers. Typical of high scores in behaviours they reported were 'reminders of errors', and 'given jobs outside job description' – apparently playing on the inexperienced individual. In contrast, respondents over 55 years of age scored highly on 'hinting at quitting', which is again unsurprising.

   Important differences were also found between ethnic groups. However, in this case it is important to warn the reader against drawing too strong conclusions as the number people from ethnic minorities was small. Without exception, ethnic groups achieved the highest score for all acts measured. The discrepancy between the 'white' group and ethnic minorities was particularly strong on behaviours of an insulting or excluding nature. Asian respondents reported the most negative acts in general, particularly derogatory behaviours. A similar picture emerged for the Afro-Caribbean participants. Chinese respondents reported most negative acts when it came to social exclusion, e.g. being ignored.

Re-estimating the numbers: how many are actually bullied?

We started this chapter by presenting some data with regard to how many people are bullied at work. We also made it clear that the available data was difficult to compare as it was often derived from different ways of measuring the phenomenon. Based on a method which focuses on the perception of the individual as to whether or not they have been bullied, and given a definition of bullying, the UMIST study gave an estimate of 10.6 per cent. The UNISON study had put a similar figure of 14 per cent. However, up to 38 per cent of respondents in the UMIST study reported to have experienced at least one negative act on a weekly or more frequent basis within the last six months and might be considered bullied. The figures are shown in Table 2.6

Table 2.6 The experience of weekly acts and labelling as bullied

Number of negative acts experienced weekly or daily	All respondents (%)	Have you been bullied? NO (%)	Have you been bullied? YES (%)
1	38	33	79
2	23	18	68
3	16	11	57
5	8	4	39
10 or more	2	0.5	13

Source: Hoel and Cooper 2000.

Once again we find that only around half those who experience negative behaviours also label themselves as bullied. This highlights the need for much more research: are those who do not label themselves doing so from ignorance of the term, or is it that their exposure to negative acts has not become sufficiently uncomfortable enough yet? Perhaps they are fighting back, and so see themselves in a conflict rather than a bullying position? These issues are high on our research agenda.

Our research often raises as many questions as it answers. The overlap between the experience of negative acts and whether or not one labels oneself as bullied is a good example this. Surveys will not answer this question, but there is a need to investigate all the different players within the spectrum, not just the bullied. We will be doing this in Chapters 4 and 5, but first we must stay with the targets and examine the effect of bullying at work.

3 Outcomes of bullying

In the previous chapter we painted a picture of bullying in the workplace by providing an idea of the extent of the problem and the very broad range of people it appears to affect. We showed that bullying is a phenomenon directly linked to people's experience of a variety of negative behaviour at work and that it is pervasive. Using the results of recent research we showed that the number of people who reported regular exposure to negative acts at work far exceeds those who labelled themselves as 'being bullied'. Whether the experience of all of these people should be thought of as bullying is an open and current question. What is beyond doubt is that a large proportion of the workforce is faced consistently with behaviour which may be construed as bullying and which may damage them.

In this chapter we will continue our journey by exploring how bullying affects people and their organisations. We will draw on research from Europe and North America. Initially we will focus on how bullying affects individuals' health and wellbeing including the most severe long-term effect, 'Post-Traumatic Stress Disorder' (PTSD). We will also examine how people re-assess their own job and their organisation (including their commitment to it), and how these effects extend beyond those directly bullied to the witnesses. Finally, we will assess some issues connected to direct costs such as health-related absenteeism, performance and productivity and exit rates.

Costs to the individual

A few words of warning are needed before we start our investigation. The relationship between bullying and health, or between bullying and any other possible outcome for that matter, is not straightforward. Most of our knowledge of the potential health effects of bullying has emerged from questionnaire-based surveys. In such studies it is impossible to be certain about the real cause and effect relationship between two variables, in this case between bullying and health. Often we can look at an *association*, rather than be clear as to cause and effect.

Box 3.1

I retired from work on grounds of ill health with a diagnosis of post-traumatic stress disorder. The total economic costs of my situation must have cost a fortune. There were costs due to sick-leave, an irritable bowl syndrome investigation, visits to my local GP, counselling by a clinical psychologist and eventually, psychiatric outpatient care.

Bill, civil servant.

Bullying affects health

A number of studies have confirmed that a relationship exists between the experience of bullying and impaired health. A large-scale Norwegian study of trade union members within a number of occupations and industries concluded that anxiety, depression and aggression were the effects most strongly related to bullying (Einarsen *et al.* 1994a). In a Finnish study of university employees, the following symptoms were reported as common amongst targets: insomnia, nervous symptoms, melancholy, apathy, socio-phobia and lack of concentration (Bjorkqvist *et al.* 1994). Based on the results of a national study of bullying in Sweden, Heinz Leymann concluded that the strongest difference between bullied and non-bullied respondents lay in 'cognitive effects' (concentration problems, insecurity, lack of initiative and irritability) and psychosomatic symptoms (stomach upset, nausea and muscular aches) (Hoel 1997).[1]

Until recently, research into health outcomes of bullying in the UK has been relatively sparse. Early studies found some links that were worth pursuing, with three-quarters of those who were currently

---

[1] Hoel (1997) provides a useful overview of the Scandinavian literature in English.

bullied reporting some damage to their health, e.g. stress, depression and lowered self-confidence (UNISON 1997, 2000). A recent British study by Quine in an NHS community health care trust found that those who had been exposed persistently to 'bullying' behaviours were more likely to suffer from stress, anxiety and depression than those who had not (Quine 1999).

Hoel and Cooper, in their national study, used two measures for health, one related to physical health: the Occupational Stress Indicator (Cooper *et al.* 1988) and the other related to 'normal' mental health: the General Health Questionnaire (Goldberg 1978). Both questionnaires have been used in numerous studies and are recognised as reliable measures of health.

For this analysis, respondents were divided into four groups:

- those who reported themselves as being bullied within the last six months (currently bullied);
- those who had been bullied within the last five years but were not being bullied at present (previously bullied);
- those who had witnessed bullying but had not been targeted themselves (witnessed bullying);
- those who had neither been bullied nor had witnessed bullying taking place within the last five years (neither bullied nor witnessed bullying).

These groups were then compared by using the results of the two health measures and the results one would expect from a 'normal' British working population.

The results showed a significant association between bullying and both mental and physical health. Compared with the 'norms' one would expect from the general population, we find very much higher levels of mental and physical ill-health for the 'currently bullied' group than for any other group (see Table 3.1). As far as mental health is concerned, the average value is well above a threshold level (see Note 1 with Table 3.1). This suggests that these individuals may benefit from screening by a mental health professional. We must emphasise that when we use the term mental ill-health, we refer to ill-health among the normal population at large and not what may be referred to as pathological mental disturbance. The second most affected group turned out to be 'previously bullied'. This suggests that effects of bullying may last longer than the direct experience of bullying.

Table 3.1 The association between bullying and health (mean scores)

Item	British scores	Currently bullied	Previously bullied	Witnessed bullying	Neither witnessed nor bullied	Prob-ablity
Mental health – GHQ score	3–4[1]	5.61	3.73	2.80	2.23	<0.001
Physical health – OSI score	30.64 (average)	41.70	35.99	32.70	30.23	<0.001

Note: [1]The GHQ score is only used for interpretation when it is high. Scores beyond a threshold mean that respondents may benefit from seeing a mental health professional. Many studies see the threshold as scores beyond 4, and some use scores at 3 or above. Remember this measure is sensitive to 'normal' mental health issues such as depression and is not designed for use with serious mental health disorders.
Source: Hoel and Cooper 2000.

Gender differences

Several studies of bullying (undertaken outside the UK) have suggested that women may be more negatively affected than men by bullying. An Austrian study found that women reported more psychosomatic complaints as well as anxiety than men (Niedl 1996). British findings have confirmed this with a stronger relationship between reports of mental ill-health symptoms and bullying for women than for men (Hoel and Cooper 2000). A possible explanation may be that women's experience of bullying in general is different and possibly more severe, independent of the number of negative acts to which they are exposed. The fact that women often work in different jobs, occupations and roles to men may account for such a view. However, there is substantial evidence that women in general report more health complaints than men. Several explanations have been suggested to account for this (Verbrugge 1985). Women are better reporters of health problems due to factors related to memory and individuals' willingness to report; working women may face additional domestic pressures, which cumulatively could affect their health.

Another plausible explanation is that men may have a higher threshold for complaints than women, linked to factors such as body awareness, sex-roles and identity (Verbrugge 1985). However, these arguments regarding men are somewhat undermined by the UMIST findings which showed that in the case of physical health, as opposed to mental health, men actually reported higher levels of physical ailments than women. Again, we need further research to explain these findings.

Bullying and post-traumatic stress

> My life is devastated, the effect of the condition denies commitments of any kind, confidence is zero, as is self-esteem. Motivation and will are a much reduced feature of life. I am void of physical, mental, social and emotional stamina. There are frequent periods of utter misery and black moods. The continuing effect of restless sleep, early waking and nightmares create feelings of guilt and anxiety. Preoccupation with and flashbacks of the experience create a deeply negative perspective. I am daily living in fear of a crisis occurring that I simply won't be able to cope with. Panic, dry mouth, aching limbs, tremors and palpitations are frequent.
>
> (John, 51, sales manager)

It is probably not difficult to gain agreement that being bullied or persistently exposed to negative behaviour at work will have some impact on health. But there is a considerable distance between relatively mild negative health effects and John's condition described in the previous paragraph. The type of report given by John pervades studies of bullying at work (UNISON 1997; Savva and Alexandrou 1998; Liefooghe and Olafsson 1999) and John is far from alone in the extreme terms he uses, requiring us to search for an explanation.

For a number of years it has been well known that exposure to an extreme event, such as involvement in an accident or disaster, can result in severe long-term health impairment. In the medical literature this type of reaction (which manifests physical as well as psychological symptoms) has been given the name Post-Traumatic Stress Disorder (PTSD). The medical diagnosis of PTSD is linked to symptoms such as vivid re-living of the event or frequent flashbacks and a tendency to avoid any stimuli linked to or associated with the traumatic event. Another common symptom observed in many sufferers of PTSD is a persistent feeling of irritability which tends to be sustained over very long periods (Scott and Stradling 1994).

Until recently, PTSD was seen to be linked exclusively to single extreme traumatic events. However, from the mid-1990s researchers of workplace bullying in Scandinavia saw a similarity in the symptoms of those who had suffered severe forms of bullying at work with those identified as having PTSD. For example, some victims reported how they started sweating and even felt physically sick on seeing their workplace and this bears a close resemblance to a common symptom of PTSD.

Based on work at a Swedish rehabilitation clinic for victims of workplace bullying, Leymann concluded that a large majority of victims qualified for the PTSD diagnosis (Leymann 1996). He also noted that many of his clients had undergone what he referred to as 'characterological change'. In other words, their personality and character had been distorted as a consequence of their experience. As a result, the person was not the same person that they had been prior to the series of traumatic bullying events.

This idea that PTSD may be brought on by long-term exposure to small traumatic events, and not just as a result of a single traumatic incidence, reflected British research at the same time. Their focus was not on bullying as such, but on the effects of persistent long-term exposure to work-related stress (Scott and Stradling 1994). They preferred to use the term Prolonged Duress Stress Disorder (PDSD) to be able to distinguish this long-term (drip, drip) experience from the acute traumatic experience identified with PTSD (Scott and Stradling 1994). Interestingly, they also came to the conclusion that people experiencing severe prolonged stress may undergo a personality change.

Box 3.2

I have been very stressed since November last year, suffering from insomnia, migraine attacks, intermittent diarrhoea, and a loud ringing noise in my left ear. When I visited my GP in December he diagnosed stress, anxiety and depression due to my work situation.

Carole, retail area manager.

A recent Norwegian study of sufferers of PTSD (or PDSD) caused by severe experience of bullying at work threw further light on the dramatic health effects often observed in individuals who qualify for the PTSD diagnosis. This study compared levels of PTSD symptoms between victims of well-known disasters and refugees from war-zones, and a large group of long-term sufferers of workplace bullying (Einarsen and Matthiesen 1999). The researchers concluded that more than three in four victims of severe bullying qualified for the PTSD diagnosis. Moreover, compared with those suffering from acute stress experiences, the bullying victims were found to have far higher levels of PTSD scores. So, how can we explain such extreme consequences?

Being victimised

To explain the experience of people who have a severe degree of reaction to workplace bullying, we will refer extensively to a report on the long-term effects of bullying produced recently by Norwegian researchers (Einarsen *et al.* 1999). In their attempt to explain these health problems, they apply a model of traumatic life events introduced by Janoff-Bulman (1992).

According to Janoff-Bulman, an event becomes traumatic when it challenges the following three fundamental assumption we hold of the world: (1) the world as benevolent, (2) the world as meaningful and, (3) the self as worthy. Janoff-Bulman postulates that every human from childhood gradually puts together their view of the world, and that in most cases this picture is positive. Having our world-picture confirmed tends to provide us with a sense of security. Of course, from time to time we realise that our picture is somewhat out of touch with the world and, therefore in need of some adjustment. By contrast, traumatic events provide situations which challenge our basic assumptions of the world to the extent that our most strongly held beliefs of the world collapse. In other words, our well-established world-picture is no longer valid.

How does this model apply to bullying at work? From believing that most people at work wish you well (the word 'benevolent'), targets of bullying find themselves in a situation where they feel that somebody may be out to 'get them'. This feeling of being 'hounded' may be accompanied by feeling let down by their colleagues and friends. For most people, such an apparent lack of support in times of great need is extremely painful. In such situations, victims may seek meaning by asking themselves the question 'Why did this happen to me?' (the word 'meaningful'). Individuals may now come to the conclusion that the world which they previously saw as positive now turns out to be neither fair nor meaningful. Victims who have invested greatly in their work and given a lot to the organisation find it hard to conclude that the world is anything but meaningful in the sense that they used to think of it. From having a good standing in the organisation, victims may now see themselves thought of by others in a negative way and their ability and use to the organisation questioned. Faced with such a situation it is not surprising that victims may come to feel that a serious injustice has been done to them and perhaps further, that the world is conspiring against them. How can it be that someone who has always tried to do their best, and was considered to be both successful and popular, now is being exposed to an relentless stream of negative behaviours (Einarsen *et al.* 1999)?

But it can get worse. Consider this scenario described in many interviews with targets in Scandinavia. Having turned to colleagues and friends for support, some of them seem to question the target's views and version of events. This might come from the questions people ask, or it may be that people's actions lead targets to infer that their case is not believed. Leymann (1990) refers to this as 'secondary bullying'.

Box 3.3

After a series of instances everyone seemed to block themselves off from me. If I started talking to someone, they would immediately turn around and start chatting with another person. If I asked a question they would pretend they didn't hear me or pick up the phone leaving me in mid flow.

Keri, voluntary worker.

Leymann compared these experiences to those of rape victims where it is common for their explanations and intentions to be questioned (Leymann 1990). In such situations, nothing appears to be more vital to the victims than trying to clear their name and ensuring that justice is done. Faced with feeling undermined on all fronts, victims will do what they can to hold on to a faltering self-esteem (Hoel *et al.* 1999). Therefore, we should not be surprised if targets who need to protect their self-image may go overboard in portraying themselves as solid, reliable and hardworking employees.

The third Janoff-Bulman concept of 'the self as worthy' suggests that a decent human being deserves to be treated well. This assumption goes hand-in-hand with a perspective of the world which suggests that as individuals we have a degree of control over the results of our actions and that normally our actions will turn out for the best. For victims of bullying, neither of these assumptions turns out to be true. For many, loss of self-respect is often inevitable in such a situation. The fact that work increasingly has become more vital to our personal identity also tends to make the situation worse. For some victims the question mark over their own self-worth may give rise to suicidal thoughts and desperate action (Einarsen and Matthiesen 1999; see also Hoel and Cooper 2000a).

Bullying and suicide

Some victims of bullying pay the ultimate price by taking their own lives (Leymann 1996). There have been some attempts to estimate the number of suicides linked to workplace bullying but these attempts have been speculative as few people take their own lives for one single reason (Hawton 1987). This should not be taken as an indication that we think the numbers insubstantial or not worth worrying about. On the contrary, the seriousness of the problem is clearly expressed in one Norwegian study which concluded that as many as 40 per cent of the most frequently bullied victims admitted to having contemplated suicide at some stage (Einarsen *et al.* 1994a). No other studies have asked directly about suicidal thoughts, although several people in the second UNISON study reported suicidal thoughts (UNISON 2000). It should be emphasised that suicide has not only been linked to targets of bullying, but also to those accused of bullying (O'Moore 1996).

Behavioural implications of bullying

In Chapter 2 we discussed what likely action targets may take when faced with bullying, where they looked to for support and how they coped with the situation. The time has now come to focus on how bullying may produce changes in the behaviour of targets and the effects this may have on the organisation.

We must be careful to avoid stereotyping, as individuals will vary in their reaction. In an interesting Austrian study of victims in a rehabilitation clinic for targets of workplace bullying, Niedl (1996) tried to reconstruct how individuals responded to their ordeal. He concluded that none of the victims resorted to a simple 'fight' or 'flight' response, which one might anticipate. Instead, he found a complex pattern suggesting that victims tried out different 'constructive' coping strategies before eventually withdrawing their commitment to the organisation.

Based on a model by Withey and Cooper (1989), Niedl mapped the pattern of behaviours of each individual victim according to four different behavioural strategies: exit, voice, loyalty and neglect (see Figure 3.1). In some cases the target had initially responded to their treatment by raising their concern within the organisation (voice). A typical follow-up strategy was to demonstrate commitment to the organisation, for example by increasing their effort (loyalty). If still no constructive results emerged, reducing their commitment to and interest in the organisation may follow (neglect). For some 'neglect' followed 'voice' directly; for others a second attempt to voice their grievance was made. Others replaced an unsuccessful attempt at bringing their case to the attention of the organisation ('voice') with 'neglect' whilst others took the final step to leave the organisation altogether ('exit') after exhausting all other feasible options.

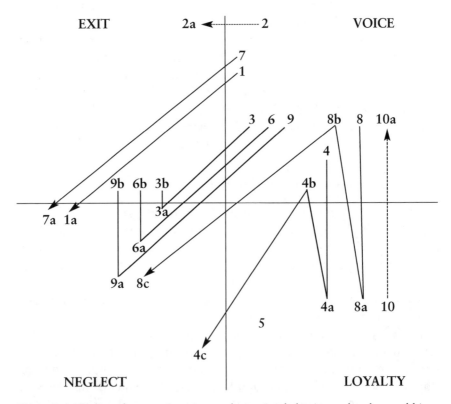

Figure 3.1 Findings from an Austrian study: coping behaviour related to mobbing, structured by the EVLN-model for the 'in-patient-group' (*N*=10).
Source: Niedl 1996.

Any attempt to piece together the past with hindsight is fraught with problems and, whilst we should be careful from drawing too firm a conclusion regarding this study, it suggests that victims are a mixed group of people who make use of different strategies in response to bullying, dependent upon the situation and their personal preferences. Even in this very small sample, Niedl demonstrates how difficult it is to find standardised behaviour patterns. In his analysis he shows that people do not leave their employer easily and that warning is given well in advance.

The following quotation from Field (1996) nicely captures some of the behavioural responses commonly observed in people who are bullied:

> The person becomes withdrawn, reluctant to communicate for further criticism; this results in accusations of 'withdrawal', 'sullenness', 'not co-operating or communicating', 'lack of team spirit', etc. Dependence on alcohol, or other substances leads to impoverished performance, poor concentration and failing memory, which brings accusations of 'poor performance'.
>
> (Field 1996: 128)

According to Field, withdrawal should not be perceived necessarily as a reflection of the withdrawal of one's commitment to the organisation. Even though people often do leave their organisation (as will be shown later), the situation facing many targets may force them into a form of paralysis.

There is a danger of a 'Catch 22' situation developing for the target. If one challenges the situation one may be in danger of acting 'over the top', and being seen as a difficult person who got what they deserved in the first place. However, the opposite strategy of 'lying low' in order to avoid any confrontation may further jeopardise one's position by leading to a situation where one is labelled 'uncommitted'. Either pattern could fall into what Leymann (1990) refers to as a 'vicious circle'. This process describes how the behavioural effects resulting from bullying make the target more vulnerable to further victimisation.

If the target becomes increasingly obsessed with their case (as they seek fruitlessly for meaning and resolution), their behaviour may gradually chip away any support they may have had at the early stages of the conflict, with increasing isolation as a frequent outcome. Einarsen and his colleagues observed situations where

targets were unable to accept colleagues holding a 'neutral' position. Instead they judged their colleagues to be 'against them',
which tended to push the targets into further into isolation
(Einarsen *et al.* 1994a).

Bullying affects the organisation

It would be naïve to believe that the negative outcomes of bullying
are limited to those directly involved. Since, in most cases, both
target and perpetrator are members of the same organisation, it is
difficult to see how the organisation could escape undamaged. Even
in the few cases where the organisation, strictly speaking, may be
blameless for the unfolding events, the bullying process is likely to
have some domino or ripple effects beyond targets and perpetrators.
It is to these effects that we will now turn.

Taking time off work

It would be understandable if targets of bullying stayed away from
work as a result of health problems or as a way of severing contact
with the workplace and the 'bullies'. However, only a few bullying
studies have been able to find a link between exposure to abusive
and bullying behaviour and increased sickness absenteeism, and in
all cases the relationship was weak (e.g. Price Spratlen 1995). In the
1997 UNISON study, less than a third of those who had been
bullied said that they had taken time off due to bullying, and most
of those only took a few days. The explanation given was that
targets were too anxious about their job to take time off (Rayner
1998). The recent UMIST study extended the enquiry regarding
time off to the witnesses and the previously bullied. A weak, but
significant, correlation was found between absenteeism and the four
groups of people in the study who differed in their experience of
bullying.

People who were 'currently bullied' had the highest number of
days off followed by those who were 'bullied in the past' and
'witnessing bullying only'. Respondents who had been neither
bullied nor observed any bullying taking place within the last five
years had taken the fewest days off. On average, respondents who
were currently bullied had taken off three and a half days more in
the last six months than those who were neither bullied nor had
witnessed bullying. This is shown in Figure 3.2. The inclusion of the

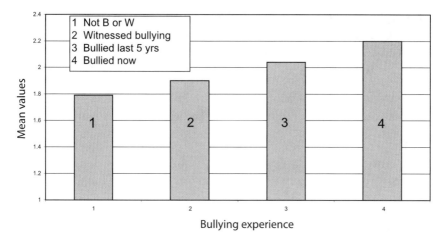

Figure 3.2 Self-reported absenteeism.
Source: Hoel and Cooper 2000.
Note: Mean values on relative scale – high=more time off.

'never witnessed and never bullied group' was useful as it shows low reporting generally.

Why is absenteeism so low? It is possible that these self-reports of taking time off for sickness are not very reliable and that this is the sort of data one should collect through means other than a questionnaire. As originally suggested in the UNISON study, it is possible that targets of bullying, in spite of growing health problems, decide to remain working for fear of a further escalation of the conflict and additional retributions if they took time off. In other words, taking time off from work may be seen as increasing rather than decreasing the likelihood of future attacks as well as undermining one's organisational standing (Hoel, Cooper and Faragher 2000). This is reinforced by the UNISON study where 'Afraid to take sick leave' attracted a 31 per cent agreement. It is possible that we are witnessing the opposite response to absenteeism, namely 'presenteeism', as a response to bullying. Such a strategy may also be seen as an attempt by targets to avoid being branded as someone who is not 'up to their job' (Adams 1992).

The nature of the survey sample may be another explanation. The most strongly affected individuals may have already left the organisation, or be on long-term sick leave and prove difficult for this type of study to reach. In addition, people who received the questionnaire may have rejected taking part as the invitation to participate in the study came from the organisation with which they may be in conflict. Finally, they may simply be too ill to take part in the study. Some comments received by Hoel and Cooper seemed to confirm that this may be the case. One of the respondents wrote:

> I cannot answer this as I have not been to work for over a year because of bullying and have only been out of hospital because of it for 12 weeks.

In another case an incomplete questionnaire was returned together with an anonymous letter from which the final passage is taken:

> I am on strong medication and see the psychiatrist on a regular basis at the nearby Psychiatric Unit. My quality of life is affected as I have so many symptoms and side effects due to unhappy work conditions causing stress.
> PS: I am 49 and should be at an age to relax and enjoy life, but sorry, I am a wreck.

If the state of a victim is such that any mention of the workplace may make them physically ill, then they are also unlikely to be prepared to answer a questionnaire which is sent to them by their employer. These examples suggest that the voice of the most severely affected victims of bullying may not be heard in studies of this type.

Exit

Considerable evidence shows that people leave their jobs as a response to contact with bullying. The UNISON surveys found that of those previously bullied, about a quarter left their job (Rayner 1998) and this UK figure has been confirmed by other studies (e.g. Savva and Alexandrou 1998). People leaving as a result of being bullied represents a very high cost to the organisation. In addition over 20 per cent of those who had witnessed bullying chose to leave their jobs as a result of their contact with bullying. These are helpful figures that provide us with data on which to build a substantial case that bullying represents a source of financial waste to the organisation.

Surveys on bullying at work usually ask those who are being bullied about their intentions for taking action. Typically, around a third of people state that they intend to do nothing and a third of people intend to leave the organisation (Rayner 1998; Quine 1999). It would appear that not all the people who intend to leave the organisation actually do. A recent American survey found a much stronger relationship between bullying and 'intention to quit' than between bullying and 'looking for a new job' (Keashly and Jagatic 2000). Clearly further case studies need to be undertaken in order to understand the dynamics.

Some targets of bullying seek an apology or redress for their experience, and may stay on in the organisation to try and achieve this. As a result they may get entangled in a long-term conflict within their organisation which can be destructive to all concerned (Ishmael 1999). The fight for redress can extend the trauma. One wonders whether this is the reason why so many people just leave their jobs – it could be the most effective way to get on with one's life and leave the bullying behind.

Work performance and productivity

From what we know so far it is unsurprising to hear that productivity is likely to suffer as a result of bullying. For example, 27 per cent of the people who took part in a Norwegian survey claimed that bullying had affected productivity negatively within their organisation (Einarsen *et al.* 1994a). In a British study, targets of bullying rated their own performance to be around 85 per cent of normal capacity. This can be compared to people who had no contact with bullying who reported 92 per cent capacity (Hoel *et al.* 2001).

However, productivity incorporates more than just a quantification of effort. Creativity and innovation are other aspects of productivity, and they are at least as likely to be affected. Creativity naturally involves taking risks, which might seem a dangerous strategy for a target of bullying, as a higher profile could make them feel more vulnerable to being bullied. It is possible that the more creativity a job requires, the greater may be the effect on productivity for targets of bullying.

Efficiency, motivation and satisfaction

According to Table 3.2, many respondents to the UMIST study thought that declining efficiency was an outcome strongly related to bullying.

Table 3.2 The effects of bullying on targets

Statement	Totally agree (%)	Partially agree (%)	Partially disagree (%)	Totally disagree (%)
Bullying at my workplace reduces our efficiency	19	14	7	60
Bullying at work affects my satisfaction	16	11	6	67
Bullying at my workplace reduces my/our motivation	15	14	7	65

Source: Hoel and Cooper 2000.

Table 3.2 supports previous evidence from the NHS in the UK (Quine 1999) and also Scandinavian research (e.g. Einarsen 1996). Indeed, it would be surprising if a target's job satisfaction were to emerge undamaged when they experience negative treatment.

The 'ripple effect'

We have presented evidence which strongly suggests that witnesses of bullying may be negatively affected. Such a 'ripple effect' was detected in the UNISON survey which showed many witnesses (22 per cent) leaving their jobs, and feeling stressed (70 per cent) by their contact with bullying at work.

We have used evidence from British studies to show how the 'ripple' affects witnesses and those who have been previously bullied. However, we have not brought into our picture the effect on those who share their life with people involved in bullying at work such as partners, friends and family. Anecdotal evidence (Adams 1992) suggests that this maybe dramatic. It is possible that the side-effects on home and social life for those involved may have an exacerbating effect on the work situation.

Box 3.4

This is the first job my daughter is holding down and she is worried sick about the potential effect it may have on her future career. Imagine seeing your once confident daughter reduced to a nervous wreck at the hand of a bully. I cannot express with words the anger I feel against this person.

Harold (parent).

Certainly, people are aware of bullying around them in general. Often we cannot pin-point where this knowledge comes from (Lewis 1999) and it is an area that needs further research. People have negative opinions about bullying, perhaps through being part of a 'ripple'. For example, 95 per cent of the 1,137 respondents to the UNISON survey (including one-third who had no direct contact with bullying whatsoever) identified 'because the bully can get away with it' as a major cause of bullying. This must imply a degree of futility for anyone at the receiving end of bullying. Most (84 per cent) of the 'currently bullied' also reported that 'the bully has done this before', and that 'management knew about it' (73 per cent) which is indeed worrying (Rayner 1998). The point is not whether the bully had done this before or whether management knew about it, but that people believed this to be the case.

Who pays for the effects of bullying at work?

There is no doubt that the biggest loser in any bullying scenario is the target. In some cases they lose everything, their job, their health, their home, their self-respect and sometimes even their sanity. The 'ripple effect' discussion points to the need to extend our investigation to those around the bully. Partners, friends and family members of the target and also the witnesses of bullying may pay a price.

But there are other losers as well, specifically employers, and unnecessary costs to the economy generally. Our figures on absenteeism related to bullying at work seem very low, but even so they add up to wastage. Based on their findings, Hoel and Cooper estimated the number of days lost from bullying. By comparing absenteeism figures, they found that 'currently bullied' people take on average seven days more off per year than those who were 'neither bullied nor witnessed bullying'. With a workforce of 24 million work-active people this might contribute to the loss of an additional 18 million working days annually in the UK (Hoel *et al.* 2001).

However, the costs of premature exit may represent a far greater cost which on its own may be massive (Rayner and Cooper 1997; Rayner 2000b). Replacement costs, including the recruitment and training of new employees, take a large slice of management time. Such time is rarely allocated to budget headings, but rather is absorbed in amorphous 'other duties' for managers and as such, lies largely unaccounted for. Most managers can think of better things to do with their time than replace employees who were lost due to contact with bullying. Moving people involved in bullying within the organisation is a further cost.

Unfortunately, the organisation which decides to take seriously the issue of bullying may witness a short-term up-surge in complaints and grievances as organisational members become aware of the possibility of seeking redress for what is considered bullying and unfair treatment. Internal investigations are notoriously costly, although such expenses do not appear in any set of accounts.

Box 3.5

I have been on incapacity benefit for the last six years. I was a skilled engineer and took home a nice pay packet every week. If it wasn't for the bullying, I would have continued paying my tax and contributed to the economy as a normal consumer.

Frank, post office worker.

Such costs are still minimal compared to the potential cost of expensive tribunal and court cases where litigation claims may lead to pay-outs or expensive early retirement deals for those involved. The publicised cases may also turn out to be disastrous in public relations terms, thereby adding further to the bill. By contrast, whilst investigation and grievances may initially be considered an additional cost, they may turn out to represent a saving in the long run, when employees are likely to become better disposed to an organisation which shows itself committed to fairness and reasonableness.

Costs to society

As with any other social problem, ultimately society will pick up the bill. The number of people who become unemployable and/or develop severe problems as a result of bullying will drain disability benefit budgets. Many targets are in the most productive phase of their life and this could have wider economic implications as their education, experience, skills and talents are wasted. If we move beyond the workplace, we will also acknowledge that bullying takes its toll on relationships and family life, realities which also need to be brought into the equation.

In this chapter we have seen how the effects of bullying can be dramatic. Although we need to do more research on the paths that bullying at work can take, we have some data. Some people, when they are bullied themselves or witness it, will leave the organisation, causing a cost to that organisation in replacing them. Other people will stay in the organisation and take a cost to themselves in terms of psychological health and work relationships. We have yet to determine the effects on those who are personally close to people who are in contact with bullying such as family and friends.

But what about the bullies? Surely if we could understand them better we might be able to ameliorate the problem either by not hiring people who will bully, or by making sure we monitor so that we can catch situations quickly and intervene? This is the topic of our next chapter, as we look at those who appear to be the 'perpetrators' of negative behaviours at work.

4   Who bullies?

Who are the bullies? Studying the 'bullies' is an extremely new area for research, and consequently the data here will be limited. This chapter will draw on the large studies of bullying at work that have been examined already in order to add to our knowledge of bullies. However, these studies have given us information from the *target's* point of view – we are not hearing *directly* from the bullies. The absence of data from bullies means that we have cast the net wider, and so in this chapter we will include evidence from other areas of study which relate to the contemporary topic of bullying at work.

The authors believe that 'bullying' has existed for as long as people have exerted power over each other. In the introduction to their seminal work on Machiavellianism, Christie and Geis (1970) describe how their research team looked for, and found, many instances throughout history (and from many cultures) of bullying tactics being used by 'great' leaders.[1] Why then, when one takes up a modern textbook on management, is bullying absent from the index? This book is not an historical or sociological commentary, so we will draw from those texts that are useful in piecing together our present-day picture.

It is also necessary to be mindful of the context which we are addressing. We are concerned with ordinary shop-floor situations, regular office circumstances and people selling shoes, carpets and financial products – not military combat. Increasingly, it has been

---

[1] Studies in Machiavellianism appear to have found a construct with which to measure people (the Mach4) (Christie *et al.* 1970). People who score highly on the Mach4 scale exhibit many behaviours one might associate with bullies, but, crucially, they are not more punitive than those who score low on the scale. For this reason the authors consider that the contemporary term 'bully' is not synonymous with 'machiavellian' in the academic sense.

common to find the teachings of battlefield commanders on the business literature bookshop shelves. Whilst some managers and organisational leaders may feel that they are in a 'war-zone', there are many differences between the armed forces and other work environments. For example, rarely is there the obedience and cohesion of purpose amongst staff in office contexts that typifies personnel engaged in armed conflict. Readers of such 'management' books may have a nasty shock when trying to transplant ideas and concepts from war conflict situations into circumstances where life or death are not the issues!

This chapter will continue to recognise that bullying is negative in nature and needs to be dealt with. However, the authors will not join a campaign which seeks to demonise 'bullies' and send them to an unpopulated island! Rather, we will take a constructive approach which is evidence-based and which will inform potential actions designed to minimise bullying at work.

Defining a 'bully'

First we must address the issue of how to study the bullies as this will help us examine existing data. How does one decide that some-one is a bully, and furthermore, how does one access them in order gather reliable information? We have already seen how difficult it is to define bullying generally. It is just as difficult to define the people who supposedly provide the behaviour to which people react. This central problem affects any study on bullies. How far would you believe a study on bullies that had drawn its participants from an advertisement in a newspaper looking for 'bullies'? Suppose instead that a firm held an election for the 'Ten Worst Managers'. One would have to question how that election had taken place and what other thoughts might have been in the mind of the voters? In other words, we must examine the 'validity' of any data on bullies before we take it as 'fact'.

Undoubtedly the search for a profile of a 'typical bully' is a tempting path. The benefits of such an approach could be the ability to screen out such people at the interview stage of selection and therefore protect the organisation from difficulty. There are some case study descriptions of bullies in the literature (e.g. Babiak 1995; Crawford 1999) and although they are often not labelled as such, we would all probably be comfortable in so labelling them.

However, these examples are very few in number and already they show a wide range of behaviours and circumstances. Those authors who have tried to 'profile' bullies (e.g. Field 1996; Marais and Herman 1997) have provided a list from which one might put together a bully profile. No one, however, has found a simple answer, leading one to conclude there must be many different 'types' of bully. Returning to the case studies then, how far can one generalise from single descriptions? When do we decide to stop inventing further 'types', and what is the value of a set of 'types'?

Marais and Herman (1997) tackled this by building on the classification of hyenas and using their behaviour as a blue-print for a typology of bullies in corporate life. Their typology showed a distinct hierarchy and described the use of different tactics by different 'types' of Corporate Hyena. The text used examples from case studies to illustrate the behaviours and their effects on targets of bullying. This text is interesting in that it postulates not only the nature of the relationship between the bullies and their targets but also the relationships between some of the bullies.

What do bullies do?

A straightforward approach would suggest that bullies are people who provide the behaviour to which our targets react negatively. Usually (but not always) bullies repeat negative behaviour towards their targets and, as such, a pattern of behaviour is established.

Managers as bullies

Respondents to questionnaires related to bullying at work in the UK and Australia have identified managers as being foremost amongst the perpetrators of such behaviour (e.g. Savva and Alexandrou 1998; McCarthy *et al.* 1996). In the UK, about 80 per

Table 4.1 Who is bullying you?

Manager(s)	Colleague(s)	Subordinate(s)	Client(s)
75%	37%	7%	8%

Source: Hoel and Cooper 2000.

cent of perpetrators include managers (e.g. Hoel and Cooper 2000; Rayner 1997; Savva and Alexandrou 1998) with a lower figure in Scandinavia of approximately 50 per cent (Leymann 1996; Einarsen 1996). It is also interesting to note that targets of negative behaviour quite often see themselves as being bullied by several people acting together. British surveys have tended to focus on the most senior person involved, but we are aware that many co-workers are also cited. This has been shown clearly in the UMIST data where respondents were asked to identify all perpetrators.

These finding should not be surprising if one takes into account that bullying is related to use or abuse of power (Adams 1992). From the discussion in Chapter 1 on features of definition we also know that, for the bullying process to develop, targets need to be in a situation where they perceive themselves as having less power than their opponents. Clearly this will be the case when the aggressor is in a senior position.

Colleagues as bullies

A substantial number of respondents identified a colleague as the perpetrator (Hoel and Cooper 2000). This was particularly true for women (40.8 per cent for women against 32.6 per cent for men). In some sectors (notably the NHS and higher education), women were bullied as frequently by their colleagues as by their managers. More than any other group, supervisors and foremen/women reported being bullied by their colleagues. This is the organisational level which possibly has suffered most from the creation of leaner organisations resulting in drastically fewer advancement opportunities. Internal competition and envy may partially explain this finding.

Subordinates as bullies

Few targets reported that they have been bullied by a subordinate in the organisation hierarchy. The study (which included many managers) found that 7 per cent of bullied respondents were being bullied by subordinates. Women were more likely to report it (Hoel and Cooper 2000), and resistance to women managers may be a possible explanation here. So whilst bullying by subordinates does happen, it would appear to be quite rare.

Box 4.1

I entered the common room at 2.00p.m. to hear Rebecca in a rage about something that I had apparently done. I heard her say: 'where is that little shit', but when she realised that I was standing in there, she lowered her voice and turned away from me and got on with some work at her desk. When I made her aware, I provoked a stream of abuse and criticism that was entirely disproportionate to what she perceived to be wrong.

Lee-Anne, lecturer.

Several bullies at once

Being bullied by a manager and colleagues at the same time appears to be relatively common (UNISON 2000). From previous studies of the problem we know that it is very difficult for players to remain neutral in cases of bullying (Einarsen 1996). It is, therefore, possible that colleagues who fear becoming targeted themselves may decide not to get involved or may be viewed by the target as taking the side of the bully. When several people are reported to be bullies, questionnaire studies have not asked whether targets think people are deliberately working together in co-operation. Anecdotal data suggests that sometimes this is the case, other times it is not (Adams 1992; Marais and Herman 1997). Indeed, intent on the part of the perpetrators of bullying behaviours has not been investigated quantitatively even though it is an important issue for targets and investigations. We also have no access to data regarding intent from people accused of bullying. One would anticipate that there would be differences in the perceptions and stories of those involved.

Cross-cultural differences

In Scandinavia, approximately the same proportion of people report to have been bullied by a manager or a colleague. In other countries, such as Australia, the United States and the UK, the proportion of managers bullying is much higher. One explanation is a low 'Power Distance' between managers and subordinates in Scandinavian countries (Hofstede 1980; Brown 1998). Power Distance is one concept used when comparing cultures and refers to people's

Table 4.2 Power Distance descriptors

High Power Distance is associated with	Low Power Distance is associated with
'Managers seen as making decisions autocratically and paternalistically' (p. 92)	'Managers seen as making decisions after consulting with subordinates' (p. 92)
Employees fear to disagree with their boss (p. 92)	'Close supervision negatively evaluated by subordinates' (p. 92)
Societal norm: 'Stress on coercive and referent power' (p. 94)	Societal norm: 'Stress on reward, legitimate and expert power' (p. 94)

Source: Descriptors extracted from *Cultures Consequences – International Differences in Work-related Values* by Geert Hofstede (1984, abridged edition, Newbury Park, Sage).

attitudes within a hierarchy. A measure has been developed and cross-cultural data has been collected for many years. Table 4.2 shows some parameters associated with high and low Power Distance scores.

In countries that show very high Power Distance, the boss is seen as correct, not to be questioned and to be obeyed. In low Power Distance countries, the boss is seen as a facilitator of the team, would expect to include the team in decision making and enter into a dialogue regarding resultant actions (Hofstede 1980). Knowing the local Power Distance enables cross-cultural workers to decide if they are going to act appropriately for the 'local' culture, their 'home' culture, or perhaps the 'corporate' culture.

Scandinavian countries are characterised by low Power Distance, and they demonstrate few psychological distinctions in their hierarchies. There is little pay difference within many hierarchies, for example. It is possible that in countries which have high Power Distance scores (such as India, Mexico and the Philippines) bullying by managers may be perceived as 'normal' and a culturally acceptable part of management behaviour.

A problem with the data

The comments in this chapter use data from surveys into bullying at work. They need to be qualified since here we run into our first problem – this data on bullies comes from the targets of negative behaviour. There are several issues that we need to bear in mind. An interesting study by Baumeister, Stillwell and Wotman (1990) examined autobiographical accounts of anger from the point of view of both perpetrator and victim. Participants were asked to describe memories of both types of situation and results showed differences depending on which role was being reported (i.e. victim or perpetrator). When respondents posed as victims, their reports were longer, more blaming and less resolved than when they posed as perpetrators. The results of this study would point to the need for care when viewing the reports from targets of bullying as the only 'truth' regarding the situation – others may have different stories to tell (Hoel *et al.* 1999).

One also has to question one aspect of the judgement of people who respond to surveys of bullying at work. For example, would one include people who 'go along' with bullying as 'bullies' or only include the instigators of the behaviour patterns – the 'ring leaders'? Furthermore, behaviour by some people appears to affect targets more than behaviour by others. An interesting study from Tasmania (Farrell 1999) asked nurses about negative treatment from different types of professionals. The results showed that although nurses received bullying behaviours from a variety of professional colleagues (such as doctors and other nurses) they were much more disturbed by behaviour which came from those within their own profession (i.e. nurses) than behaviour which emanated from others (such as doctors).

We must also take into account the position of those who are reporting the behaviour. Hoel and Cooper's study (2000) was the first to fully investigate reports from those at different hierarchical levels. The closer to the top of the hierarchy, the more likely the target is to report being bullied exclusively by one person. This finding would reflect the structure of many organisations which tend to thin out at higher levels.

These examples and subsequent discussion illustrate the complexity surrounding who might be counted as a 'bully'. We must accept such complexity, and be aware that this will affect any data about bullies which emanates from targets. Having drawn this to the reader's attention, we will now examine other data that we have found in the incidence studies which describe the bullies.

Age

In UK studies, bullies are reported as usually being older than their targets (e.g. UNISON 1997). Managers are likely to be older than subordinates and so, like gender, this just may be a function of position. Age, however, may be significant as it may be a source of power for the bullies. As we get older our confidence usually increases and perhaps we are more direct in what we say. It is also possible that, in the targets' eyes, age confers power, and so targets are less likely to challenge older people directly, possibly allowing a vicious cycle to develop.

Gender

In Britain, women are reported as bullies less than men overall. When one takes into account the fact that there are fewer women in management (there are twice as many men in management positions in the UK) this means that the numbers equate between the sexes. Such data would suggest that bullying in the UK is a function of abuse of power rather than connected to gender.

In UK studies, there is some evidence of gender difference, although whether this is a function of the bullied or the bullies needs to viewed very cautiously (e.g. UNISON 1997; Hoel and Cooper 2000). While roughly an equal number of women and men are bullied, our data does not show clear patterns of who bullies whom, as demonstrated in Table 4.3.

The Hoel and Cooper study used a wide sample across several sectors and their results may be explained by reference to characteristics of current labour markets. Many types of work are mainly female or mainly male. It is possible that women bosses are present mainly in female sectors of the economy, and simply because of this

Table 4.3 Gender in bullying. This table compares data regarding gender and bullying from two large UK studies

	Hoel & Cooper (2000)	UNISON (1997)		Hoel & Cooper (2000)	UNISON (1997)
Men bullied by other men exclusively	62%	25%	Women bullied by other women exclusively	36%	26%
Men bullied by women exclusively	9%	50%	Women bullied by men exclusively	29%	63%
Men bullied by men and women by	29%	25%	Women bullied men and women	33%	11%

are less likely to be reported by men. In contrast, the UNISON survey had a high proportion of respondents from the NHS, and almost all from the public sector. It is possible that the narrower sample can account for the different patterns where more cross-gender bullying was found.

Other sources of data regarding bullies

Incidence studies have revealed some data regarding bullies. This data has been fairly basic as it has been derived from targets of bullying who, as previously discussed, may cite certain people to be bullies more than others. There are also other reasons why we should be concerned about the validity of the data. It is therefore important to extend our search to other fields of investigation and try to see whether we can find data on bullies, even though it may not be labelled as such. This section will spread our net wider to look at parallel areas of interest, beginning with the vast area of writing on management.

Bullying in the management literature

Given that managers are frequently involved in the dynamic of bullying, it would be reasonable to expect some attention paid to it in the well-developed literature on management and management style. Although there is an established difference between management and leadership (Kotter 1990), the term 'management' will be used interchangeably with 'leadership' and we acknowledge that it will be necessary to distinguish between the two in future developments.

Bullying as a management style

The early 'management styles' literature developed in parallel with ideas on how to motivate people. That is, how a manager chooses to motivate their staff establishes a 'style'. Perhaps the most pervasive theory in this context is still that of Douglas McGregor (1960) who postulated two sets of attitudes which managers could bring to their role of managing people. His work was based on anecdotal experience over many years as a consultant. The 'Theory X' manager (as he called them) considers that the subordinate comes to work to earn a wage, needs telling what to do and can contribute little of any value to the work process (as they are not interested). In contrast, the 'Theory Y' manager starts from the belief-set that all employees want to be involved in the process of their work and can be 'empowered' in modern-day terminology. Given this set of beliefs about what motivates people, Theory Y managers would be inclusive in their management style, coaching and enabling staff to take responsibility and valuing their input to the process of work and its organisation.

Box 4.2

On returning from sick-leave I found that the telephone was removed from my desk and the filing cabinets locked and with no trace of the keys. When approaching the secretary I was told that my line-manager has informed her that I would have to ask her to use them each time and that all telephone calls should be taken from the secretary's office and in her presence. On the way home from work that day I received a call from the secretary telling me that she had been asked to tell me that my desk would have to be removed to make place for another secretary.

Michael, prison service.

Such 'schema' (as they might now be termed) regarding the belief-set of staff provide two very different starting places for the treatment of staff at work by managers. McGregor encouraged managers both to reflect on their own starting point and also to adopt a Theory Y approach. His text set the tone for later work where the emphasis is on the benefits of Theory Y. It is possible that managers who behave negatively towards their staff at work begin from a Theory X position (Ashforth 1994).

Although McGregor did not conduct empirical work to back up his ideas, many studies on his theory have been undertaken since. Yukl (1994) provides a useful review of these studies, which reveals that there is no conclusive evidence on which is the 'best' style for high productivity etc. McGregor's own emphasis on Theory Y as the 'best way' to manage staff has not been proven from a mechanistic or a productivity point of view, although some see it as highly appealing.

A recent article in the *Sloane Management Review* (Ghoshal *et al.* 1999) outlined the managerial challenges for the new millennium. It indicated that McGregor's 'Theory X' managers and the theories that preceded it (such as Taylorism), although archaic in business thinking and writing today, are still alive and well at the turn of our new century. The design of telephone call centres demonstrates this well. Here staff are guided in their work by software and almost 'managed' by their terminals which record all events providing data for later productivity bonuses – strongly reminiscent of a Taylorist approach.

Such attitudes must be re-examined if we are to move towards a balance for the employee. There will always be a strong priority for managers to achieve productivity targets, but the design of jobs and work sometimes appears to view the human as just part of the nuts and bolts of the 'productivity' machine which was mocked in Chaplin's 1936 film, *Modern Times*.

Management behaviour

Yukl's (1994) overview of leadership theory provided a reflection on many years of struggle both to conceptualise and measure management and leadership behaviour. He does attend specifically to negative behaviour when considering tactics that people use to exert influence over each other: 'The research has neglected forms of influence behaviour that involve deception, such as lying, deliberate misrepresentation of the facts, exaggeration of favourable inform-ation, and suppression of unfavourable information' (p. 237). Ashforth notices this in his introduction to the concept of 'Petty Tyranny' in managers (1994). He made the case for looking at negative behaviour *per se*, rather than just assuming that the absence of positive behaviour is negative.

Such an investigation is rather difficult to achieve in practice however. After many years of research into management style, Yukl has been clear on the key problem in management and leadership research, namely the lack of validity of measures and instruments. In other words, there is not a valid measure of management style (Yukl 1994).

Moving now to the effectiveness of leadership, one would hope that we have something to learn here from the myriad of studies. In their review of management competence and personality, Hogan *et al.* (1994) comment on the lack of data regarding incompetence

and lack of effectiveness. They further state: 'managerial incompetence [has been found] to be associated with untrustworthiness,
over control, exploitation, micro-management, irritability, unwilling-
ness to use discipline' (p. 495). Interestingly, as personality special-
ists themselves, these writers found little of conclusive value in the
literature regarding success or failure in managerial performance
and manager personality. The high level of speculation regarding
the personality of bullies among some texts on bullying would seem
at odds with this finding from the management literature.

Box 4.3

He would often refer to the commitment of members of his
team, by pointing out that these persons had not taken their
annual leave for eighteen months and that they never claimed
for expenses. When I went on holiday for a week this summer
with my wife, I was made to feel that I was not pulling in my
weight and that my absence would put further pressure on my
colleagues.

Ian, manufacturing supervisor.

Management and power

Power is a well-researched area within management literature, but
once again studies are often seeking positive ways to use power,
rather than exploring what happens when power is used negatively.
Quantitative studies rarely focus on this, and when they do,
complexity emerges with a muddled picture regarding rewards and
punishment which can be used together (e.g. Yukl and Falbe 1991).
This is helpful for those studying the bullies, as it places a warning
indicator that we are grappling with complex interactions and
dynamics which are still not understood at a basic level after many
years of research.

Tjosvold (1995) pointed out that the source of power (such as
power from status) is only one dimension in the process: 'How the
power base is used, however, could be quite critical. The same
power base can be both used to reward and punish' (p. 724). This
would be the case for a manager in position of authority. Rayner
(1999c) found that the focus for the power (i.e. undermining

someone on a personal level versus the manipulation of the ability to achieve a task) may be a further issue of relevance when studying bullying at work.

Power is used, amongst other things, to control. As a key aspect of a managerial job is control, this forms a link for us in our understanding of bullying. In some societies (for example where there is a high Power Distance) bullying may be a legitimised form of control, but in other societies, such as many in Europe, bullying lacks legitimacy as a form of control (Leymann 1996). By using control as part of our scenario we can also understand the fluidity that can typify bullying relationships. Power and control are both dependant on where one happens to be in relation to others and also on the resources we have at our disposal at any one time. As we study bullying at work further we may find that the legitimacy of power and control has considerable more salience than any issue to do with the personality or individual attributes of the bully. Bullying may reflect the moving dynamics of power (or lack of it) and achievement (or not) within the organisation. Such issues are examined when we consider organisational aspects more fully in the next chapter.

Conflict studies

An area of study related to bullying at work is the study of conflict. Historically based in the study of international conflict (e.g. the Cuba Crisis), there are clusters of researchers who have progressed to the study of conflict at an individual and a dyadic level within the context of the work environment (e.g. De Dreu and Van de Vliert 1997). Refreshingly they do not see conflict as necessarily negative (Van de Vliert and De Dreu 1994) and consider conflict 'normal'. A good example they use is when a new work group forms with quite healthy disputes (called 'storming') which enable group members to test and explore the limits of the group (Van de Vliert and De Dreu 1994).

A study by Karen Jehn (1994) found that groups of students who experienced high emotional conflict had a drop in performance. This compares with those who experienced high task conflict, which increased performance. She also measured satisfaction levels of the group members, and found lower satisfaction reports when emotional conflict was high, but neutral reports when task conflict was high. This may hold value in our search for bullies as they may be people who can touch others, emotions. Rayner's study (1999c) also found that negative bullying behaviour was clustered around task and/or personal attacks. Her results showed that people were bothered by both types of behaviour from their managers, but she did not measure the level of work satisfaction, so it is not directly comparable to the Jehn study.

Some recent work (Van de Vliert *et al.* 1995) used videos of Dutch police officers attending a training course. Included in their course was the task of persuading another person. The researchers then asked 'experts' to view the videos of the exercise and judge the tactics used by individuals during their task of persuasion.

Figure 4.1 shows the descriptors they used (which relate to conflict management) together with their relationship to effectiveness and agreeableness when they are studied as single or discreet moves or actions. This is conventionally described as a 'ladder of effectiveness' (p. 278).

This study revealed that people may use not one, but several methods of persuasion, and that the effectiveness was determined not by the absolute level of any one chosen tactic, but by the combination of different tactics. 'All this contradicts the common view that the dominant type of conflict handling behaviour is also the dominant determinant of behavioural effectiveness ... the

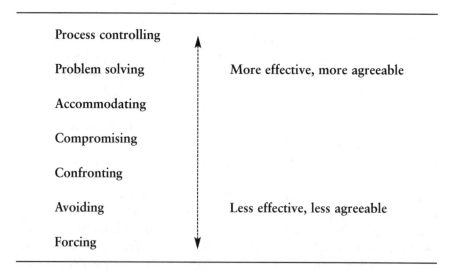

Figure 4.1 The 'ladder of effectiveness'.
Source: Van de Vliert, Euwema and Huismans 1995: 278.

gestalt of components counts, not the sum of the parts' (pp. 277–8). They found that problem-solving behaviour was 'especially effective if a superior combines it with forcing vis-à-vis a subordinate' (p. 278). So these researchers found that people were using a whole range of different behaviour even within a single incident. They also found very complex patterns. To a large extent, the setting was contrived, and did not include the cost issues involved for subordinates in permanent jobs as one would find in workplace bullying. However, even after trying to make the exercise as straightforward as possible, the researchers found a complexity that was surprising, and it challenged the simple 'ladder of effectiveness' they had been using as a model.

Complexity!

Leadership and conflict behaviour studies both appear to be converging on the need for the investigation of multiple parameters in parallel in an attempt to shed light on the complexity that is closer to real behaviour in the workplace. We will not get very far by looking for single types of behaviour that fit nicely into academic categories – as was shown when we examined the management literature. This type of evidence forces us to bring our academic studies closer to real life. It also shows the difficulty we will have in determining what constitutes a typical 'bully' and allocating their behaviour into neat categories.

Box 4.4

In recent years the staff meetings have stopped being exchanges of views but just reading out of decisions already made. And any attempt to raise any points whatsoever, were just completely ignored or swept away. And that, in addition to the fact that people were constantly being ridiculed when absent, had the effect that people stopped putting themselves in an embarrassing position of being publicly ignored. It wasn't so much overt ridicule, but covert.

Naomi, catering worker.

If we embrace the notion of many different behaviours occuring in parallel, then we can see how this will affect the task of those who conduct investigations into bullying. If the person accused also achieves well in other areas, and the target has few pieces of unambiguous evidence of bullying behaviour (as is often the case), the investigator is in a tricky situation. How far reports of negative behaviour will stack up against those accused of bullying may also relate to other elements of the accused's competence, which may also have evidence of positive behaviour.

Bullying in aggression studies

All aggression is not identical. A useful, albeit simple, dichotomy is the distinction between angry aggression and instrumental aggression. Depending on the circumstances, bullying might be seen as angry aggression such as where the 'bully' fails to control their temper and lets their emotion take over in an angry attack on the target. To the authors' knowledge, few workplace-based studies exist on angry aggression (e.g. Neuman and Baron 1998). In designing an ethical methodology one cannot put people under so much stress that they explode or shout at people in the cause of research. The previously described study into accounts of anger by Baumeister, Stillwell and Wotman (1990) comments on this point: 'One reason for the paucity of data on anger is the difficulty of studying it under controlled conditions' (p. 994).

Bullying could also be seen as instrumental aggression which is more thoughtful, principally designed to achieve an end (Hoel *et al.* 1999). This might be thought of as manipulative, with perhaps less overt anger. Unfortunately this topic has received very little atten-tion (Geen 1990) from academia, partly because of the difficulty of designing studies. However, one can see how a bully might weigh up the pro's and con's of being aggressive by bullying someone and decide to use such tactics. There may be little 'cost' if they coerce or undermine in this way, but plenty of 'benefit'!

These perspectives could prove useful when we consider what to do about bullying at work. If one comes across an 'angry' bully, then one might take measures to help them control their emotion. In contrast, where someone is being instrumental and using bullying as a means to a calculated end, then the organisation would need to change the 'cost – benefit' equation quite radically. One set of interventions would be inappropriate for the other group.

Studying aggression is fraught with methodological problems, not least ethical considerations, and studies typically are between strangers in contrived settings. Essential to the notion of workplace bullying, however is the long-term context of people who know each other, with a past and a future together and risky implications such as the loss of employment (Rayner and Hoel 1997). Finding applicability of research from aggression literature into the bullying at work context is difficult.

Psychoanalytic approaches to management

Approaches to organisation theory using psychotherapeutic tools are not common, but they do consider the negative side both of organisations (e.g. de Board 1978) and individuals who are in positions of power (e.g. Kets de Vries and Miller 1984). Most writers of this genre would see the bully as someone who is acting out problems to do with their inner psyche within the bullying relationship (e.g. Bowles 1991). For example, they might try to project their own frustrations onto others and scapegoat them (Thylefors 1987). If readers are interested in such an approach, they should read the source texts as justice cannot be done to them in a book of this kind. The target of the bullying may also be fundamental, as they may remind the bully of an unresolved relationship or elicit a memory in another way, to which the bully then reacts. The analytic approach would see the bullies as understandable, albeit deviant, and they would also suggest some types of categorisation for such people.

Stereotyping

Chasing such stereotypes can cause problems, and we are seeing this emerging in the research into school bullying at the moment. In school bullying, the perpetrators have been typically portrayed as clumsy dollards who are not very bright or well socialised. Sutton has recently completed a study (Sutton 1998) where, contrary to this stereotype, he found that many school bullies were highly skilled socially, and well liked by many class mates.

Some writers also seek to down-grade adult bullies at work, attempting to show them with flawed self-confidence (e.g. Field 1996). An article by Baumeister, Smart and Boden (1996) provides a concern that this approach maybe as irrelevant to workplace bullying as it is in schools. Their work presented a wide-ranging review of leaders, and made a strong argument that, far from lacking in self-esteem, such leaders could be judged as having high self-esteem. Their study highlighted the need for a technically correct approach. If one goes looking for psychologically 'crazed' bullies, one will find them – the investigator must be careful of their method and their own bias (Popper 1959).

Trying to crack the complexity

So far this chapter has demonstrated the complexity involved in the challenge of studying bullies. We do need data though, because we need to make appropriate, constructive interventions. As with targets of bullying, the bullies come from all occupations, and they are likely to be good at things other than bullying, such as accountancy, quality control processes, web design or theatre management for example. Ideally we need to be helping these people amend their behaviour so that their strengths can continue to be utilised within our societies.

Some commentators such as Neil Crawford (in Adams 1992; Crawford 1999) and Blake Ashforth (1994) help us in trying to pull together various aspects that may be affecting the bullies. Such aspects would include their work situation, the individual stresses they may face and their own life experiences. Both these writers go beyond the individual level of analysis (i.e. parameters relating to the bully as a person) to the work group and also to the climate in which the bullies are operating. We will be discussing these wider issues in the next chapter.

Direct attempts to study negative behaviours at work

To date, only one quantitative study has directly examined the perpetrators of negative behaviour (Rayner 1999c). This cannot be called a study of 'bullies' as at no stage were respondents asked whether they were bullied, but it is of value. Rayner followed Ashforth's idea of asking subordinates about their manager (Ashforth 1994). Ashforth had asked two subordinates to report on their (same) manager's negative behaviour and found a strong correlation between ratings of subordinates who shared the same manager. Rayner decided to take the work further by also collecting data from the managers themselves as well as the subordinates.

In this study (Rayner 1999c), all staff were asked about their experience of negative behaviours (remember no 'bullied' label was used) and the identity of their manager. She then grouped the respondents' answers by manager and examined the patterns within the work-groups. Only groups where three or more staff had responded were taken, and even this might be seen as a low number if the manager was responsible for a large group of people.

That said, patterns did emerge. Some managers were reported by the whole work-group to use much more negative behaviour than others. Other managers were reported to use high levels of negative behaviour by only one or two people in the work-group. Sub-ordinates in further groups reported consistently good treatment from their manager. Managers were labelled using internal norms (that is, people were reported as relatively tough or sweet as managers). Those who used very little negative behaviour were labelled 'angels', those who used relatively high amounts of negative behaviour towards most staff were termed ' tough managers', those who were reported very high on tough behaviour by single individuals 'victimisers' and the rest 'middlers'. Around 80 managers were classified using over 250 reports.

At this point we should note that this study falls into all the problems and traps of stereotyping that have been discussed. The managers and their behaviour have been classified in an artificial way, albeit with rules based on what we know about bullying from other studies.

The managers themselves were then tested on a personality measure (the Hogan Personality Inventory), a stress measure and a mental health measure – the latter two used the Occupational Stress Indicator (OSI). We should now pause and consider what results we might expect. Remember that we have placed managers into four categories – 'tough managers' (high levels of negative behaviour reported by many in the group); 'victimisers' (a few people reporting unusually high negative behaviour); 'middlers' (where nothing unusual was going on); and 'angels' where very low levels of negative behaviour were at work.

If we believe that bullies are psychologically disturbed in some way, we would expect some indication of this in the personality measure of the managers. In addition, we should be able to pick up unusual scores for mental health. If we think that managers are reacting to high stress themselves, then this should be reflected by tough managers scoring highly on the stress indicator. The numbers of managers who responded to these tests were small (around 30) – but this number was seen as adequate given that the scenario envisaged was that of trying to screen out bullies from a short-list of candidates applying for a job.

No differences were found between the groups. There was no significant difference in the scores on stress or mental health between tough managers, victimisers, middlers and angels, and further qualitative analysis of the scores showed nothing particularly noteworthy in the results. Qualified psychologists could not predict the group to which the personality profiles of the managers belonged. The psychologists (correctly) pointed out that the *interaction* between a manager's personality profile and their subordinates' personality types would also be important (Rayner 1999c). The Hogan group should be complimented on their willingness to participate in such a study. It is also worth noting that they have since released a 'Dark Side' inventory, the testing of which would be very useful in these conditions.

Implications for the organisation

The last study demonstrates that the idea of screening for bullies at the selection stage of recruitment is not worth pursuing. It also suggests that employers should wait for studies which use much larger samples of managers before trying to employ stress and mental health as singular measures to back up the labelling of people as bullies. In effect we have data which shows that, with small samples of managers, the situation is too complex to draw any conclusions. Practitioners should therefore be wary of those who 'profile' bullies.

The implications for investigation into bullying at work from the incidence surveys are not very encouraging either. These studies indicate many people may be involved, hence making investigations extremely difficult. Considerable amounts of evidence may be necessary before conclusions can be drawn and situations may be inherently complex. The concept that single bullies victimise single targets clearly does exist in reality. But multiple targets with multiple perpetrators may also exist just as frequently.

Indeed, how far bullies are merely conforming to a set of established behaviours themselves might be of critical importance to an industrial tribunal, for example. If one has a culture where bullying can exist and remains unchallenged by others for a considerable amount of time, the 'bullies' may see themselves as just fitting in.

Box 4.5

The few times I had a conversation with any of them I got on with, one of the ringleaders would immediately interrupt us asking the person I was talking to join them instead. In most cases they did and I was left alone.

Keri, voluntary worker.

Such comments bring us to our next chapter where we will be examining corporate culture. How does the culture of an organisation reinforce bullying at work? How can we think through the problem at the organisational level?

5 Bullying and corporate culture

Why should we examine corporate culture in a book on bullying at work? In Chapter 2 we saw how targets of bullying may spend several years experiencing behaviours at a cost to themselves, often known about by others. When reading such stories in the press, a question which most of us will ask is 'Why wasn't it stopped earlier?' The authors see the culture or climate as the major organisational element that allows bullying to continue by upholding norms of behaviour from an era when such behaviour at work was not questioned. In this chapter we will use a model called the 'culture web' to pick apart the various aspects of culture and, using examples, show how these can both support and diminish bullying at work.

There are times when bullying is identified as a very isolated dynamic between two individuals, but more often it is not. The UMIST study confirmed that the target often tells someone about their experience (Hoel and Cooper 2000). Also there are witnesses to the bullying, and frequently people are bullied as part of a group. These factors mean that bullying is not usually something which happens behind closed doors, rather to the contrary, many people know about it.

One aspect that has surfaced in our surveys on bullying is the presence of an acceptance of bullying at work. In both UNISON surveys, over 90 per cent of all respondents cited a cause of bullying as being 'bullies can get away with it' and 'workers are too scared to report it'. These findings tap into pervasive beliefs, reflecting that bullying has been around for a long time, we have all known about it (even if we haven't experienced it) and we know nothing gets done. Remember, these are beliefs – they don't need to be true, but their existence as beliefs sets up structures within which we make individual and collective sense of our world (Rayner 1998).

When trying to combat bullying at work, one stumbles across some extraordinary (and erroneous) beliefs, the origins of which will be difficult to disentangle. However, we do know that once people have beliefs, they tend to look for evidence that will reinforce their beliefs rather than change them. It is crucial therefore that all of us who work in organisations look rather carefully at our daily lives to see how we might be supporting the notion that 'bullies can get away with it' and that 'workers are too scared to report it'. Previous chapters examined bullying on the individual level and extended this examination to the work-group. We now move our focus to the full organisational view. In order to do this we will utilise the concept of organisation culture.

Corporate culture defined

What is corporate culture? Some people see corporate culture as an expression of the 'personality' of the organisation. Anthropologists have provided a base from which we can study the culture in our workplaces, and like them, we use observations of what people do, how they make decisions and their group beliefs (amongst other methods) to infer the culture that we can then express. However, it is difficult!

The briefest and pithiest definition of corporate culture is 'How we do things around here'. More complex definitions extend to teasing apart the various elements which make up culture in organisations which reflect how notoriously difficult it is to examine culture (Brown 1998). For those of us interested in measurement, culture presents a unique problem because of the many influences to its formation, how differently it can be perceived by those who experience it, and the task of separating out cause and effect. If culture is about the attitudes and beliefs of those within the organisation, we have to think about measuring those aspects.

If we think of culture as a way in which bullying is supported within an organisation, we should also be able to use culture as a way of reducing bullying. To do this we need to consider it in terms of the ways that we can change and manipulate it. If bullying is an abuse of power, how does the organisation allow situations to occur where bullying takes place and yet it is not reported or where there is a lack of action even when it is reported?

The culture web

Any good consultant will arrive early on their first visit to a new client. Ostensibly they may appear to be sitting in reception and watching the corporate world go by. In reality we all have our antennae highly sensitised in these new environments, but what we pick up and how we process information to form a 'view' of the organisation is difficult to analyse. But we all do it.

Many academics have presented ways and models of thinking about culture (e.g. Schein 1992; Johnson and Scholes 1997). Our selection of which model to use has been based on usefulness. A very practical model is the 'culture web' presented by Johnson and Scholes in their book on corporate strategy (Johnson and Scholes 1997). The 'culture web' is shown in Figure 5.1.

Rather than immediately trying to label the culture (the 'paradigm', placed in the middle of the diagram), they work from

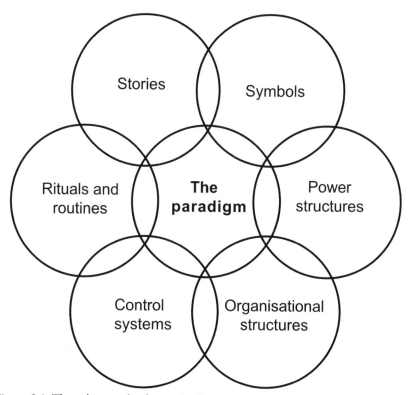

Figure 5.1 The culture web of organisation.
Source: Johnson and Scholes 1997.

the outside, that is, from those parameters making up the culture. This model illustrates the interconnectivity of the elements as well as providing a set of windows through which to structure our analysis of the organisation. We will work around the circle of the culture web and separately look at the components before we put them together and see the whole. We will illustrate our points by using real examples from workplaces but we have changed details so that the organisations cannot be identified. Readers will notice an absence of references throughout most of this chapter as we want to protect our information sources.

Stories and the culture web

Returning to our consultant – one thing they are doing is listening to the interactions of the receptionist. A good receptionist will be the hub of an organisation. What stories are being told to him or her? The stories that are told in a firm tell us about the values and beliefs of the staff because the stories are seen as worth repeating. What constitutes the 'worth' which makes them valuable to pass on? Who are the heroes and heroines in these stories? How have they earned their reputation? Do the stories revolve around positive corporate values (such as great achievements with customers) or are they less to do with corporate values and more to do with 'horror stories'? In other words, stories tell us where those in the organisation put value! When a story is passed on it is a piece of 'value'.

A certain organisation knew that they had a major problem with bullying. Many of the staff worked in a call centre environment where people answered phones taking orders for goods. Another large group of people worked in accounts, chasing up the money owed. The Director of Finance was found to be bullying her staff badly, the new bullying policy was used and she was fired.

Throughout the process of devising this particular anti-bullying policy, staff had been included. Many people knew of the policy, but it had not achieved a high profile in the culture of the organisation. It was only after the Director of Finance was fired for bullying (and the knowledge got out) that bullying made it onto the corporate culture agenda. Those in personnel reported a 'shock-wave' through the organisation as the story spread. The story communicated more about the new way of dealing with bullying than any amount of 'official' literature. For this organisation, their experience of stories worked for them and helped change their culture. The story made most staff think twice before they engaged in bullying behaviour.

In another organisation, a very senior manager was found to be a serious bully who could be dismissed in line with their policy on bullying. However, this senior manager threatened to counter-sue the firm if they fired him. Personnel were asked by the Board of Directors to estimate the costs of fighting the case. A deal was struck so that the senior manager retired early with (to him) a handsome payoff. The cost of the deal to the firm was around half of what the firm would have had to spend fighting in the courts. It seemed to those who made the decision that everyone had won – the organisation had got rid of its bully, protected its staff, saved its reputation from a nasty court case, and ended up saving money it would have otherwise had to spend on court fees.

Unfortunately, this incident got into the local press. What did the people in the organisation make of it? The story that circulated amongst the employees was that bullying was rewarded by a nice big payoff. Trust was lost. Employees didn't care about the corporate view. Personnel staff found their sessions in awareness raising on bullying shunned, and had to review completely the policy and their practice for the future. The Board genuinely did not realise the repercussions, and in retrospect would have possibly taken a different decision. In the end more money was spent trying to repair the situation than was saved.

Routines and rituals and the culture web

These reflect the heart of an organisation's values. Fundamental to their existence is people's participation, even that which may be condoning them through inaction. Here one can find acceptance of bullying enmeshed in rituals that, in the telling, reflect some very negative attitudes. One newspaper was said to have an editor who each week would pick on a staff member who had not pleased him, and for that week, that staff member had to sit at an empty desk isolated in the middle of a large room whilst other colleagues worked around the edges of the room. They had no computer, no telephone and, for a journalist, effectively no means of working. No one would talk to them for the week because they were worried about being in the 'hot seat' next! This ritual was repeated with the full knowledge of the staff and all who walked through the office. It was an accepted ritual, if an unwelcome one.

We hear of humiliating and degrading rituals. Those found in the uniformed services have attracted major press publicity in particular. Some rituals are inculcated into the culture in a way as to make us 'feel' they are not personal (as they happen to everyone). Nevertheless, they can distress and disturb. Consider the theatre nurse who left her hospital because she couldn't bear to join in the ritual of throwing extracted kidneys and other organs around the theatre in a game of 'catch' before they were put in the waste container. She felt the ritual was degrading to her own high professional standards as well as to patient care. Some staff saw it as harmless because the patient never knew about it, others positively enjoyed it as it distinguished their 'tough' team from others. The theatre nurse was accused of having 'no sense of humour' by the surgeon and 'lacking in team spirit'.

On a less dramatic level, most of us know people who always make others wait unnecessarily when they have an appointment. Conversely a senior manager in a utility industry got his secretary to advance their office clock by fifteen minutes before certain people arrived. The senior manager's first question to his visitor would be to ask for a reason for their lateness. Once the two people were safely in the senior manager's office, the secretary would have to re-set the clock to the correct time. Such office 'games' can be construed as an abuse of power. Another example is the group where the newest member has to wash all the cups after tea. Is this just an innocent time-honoured ritual or a debasing (albeit effective) reminder for the newly arrived to be careful? Why make points in this way? Since these are not personal (which is the case for a real rite or ritual) they are hard for people to label as bullying because they know that everyone has had to do them. Staff may still find it distressing however. Such small events allow the belief that staff have to do what others want of them, whether it is servicing the needs of the organisation or not (Crawford 1997).

Control systems and the culture web

We measure what we care about, and in this way we can monitor and get closer to controlling it. One way of avoiding the need to deal with bullying is not to have a policy and to ignore its existence! By excluding bullying from the 'official' dialogue within the organisation some people think that they can ignore it. Well, they probably can if there are no means of communicating officially about it, but the costs to the organisation and its staff will continue.

Many people have a good idea where bullying is occurring in organisations. Human resource staff need to know where it is going on. One financial services organisation we know of stumbled across unexpectedly bad treatment in an area through the use of good control systems. As a financial services organisation, perhaps one should not surprised that they kept good records which were regularly reviewed, as that is part of their discipline in their core business.

One set of records at this financial services company showed two middle managers resigning without new jobs and a third person in the same department taking early retirement. The average tenure amongst the three managers was over ten years, and these departures occurred within a four-month period. Investigation revealed that a new senior manager had been appointed shortly before the departure of these staff. With permission, personnel officers went to the homes of the staff who had left and undertook exit interviews to discover why they had resigned. They uncovered a barrage of negative behaviour by the new manager which the staff found very difficult to deal with and, as a consequence, they had decided to quit. The firm fired the senior manager and persuaded two of the managers to return back to work for them. As far as we know, they are still there, doing a good job. From the organisation's point of view the bullying had occurred in an unexpected area, but the good control systems allowed them to monitor the whole organisation and pick up this anomaly.

This a positive example of people who were able to construct their practices, which included control systems, to alert them to the possible existence of a problem. They took action, but wouldn't have known about it had the systems not been in place.

Control systems can come under distinct strain when an organisation attempts to deal with bullying. If people see these types of behaviour as 'normal', reports may be hard to obtain. It is also possible that those who observe or experience bullying at work fail to report it for fear of retribution. If one needs to know exactly where it is occurring, then anonymity and confidentiality can be compromised.

Inappropriate control systems can positively support bullying behaviours. Consider the highly competitive environment where group bonuses are in action – what happens to the people who are the under-performers and seen to drag down the overall performance of the group? This is not to say that less effective staff should not be dealt with, but do the control systems support the problem being addressed in a way which gives them dignity and respect?

One example of this is a garment company that introduced 'team working'. The chosen system was one where management deleted the supervisory grade, raised the production team members' wages accordingly and had the teams manage themselves. This took place with no training on interpersonal management. The 'self-managed' teams comprised of groups of about six people who had a certain number of garments to make each shift. Bonuses were paid on the number of garments produced after regular production targets had been met, and split between the team members.

The actor Arnold Schwarzenegger could have used some of the behaviour reported by these teams as inspiration for his less appealing characters. Heckling and other verbal abuse was directed at staff who weren't 'up to speed'. They were isolated during work-breaks and excluded from team decision making. They ended up in a no-win situation, were very miserable and sickness rates shot up. The personnel department was instructed not to intervene, but to let the situation settle down. Not given any access to the corporate control mechanisms, some of the teams became desperate to get rid of their less productive members. When these staff went off sick, the anger amongst the remaining team increased as the inexperienced temporary workers were even slower than their 'off-sick' colleagues and they had no hope of making the bonus. No formal control systems were in place. Some of the less productive people on the permanent team responded by taking sick-leave when they couldn't stand it any more, but the remaining staff who were trying to make their bonus were also stuck. Eventually the 'law of the jungle' took over and it was only after cars were vandalised that local personnel management insisted that further action was needed by Head Office. As a result the system was suspended.

Team working of this type can be extremely effective since it gives staff control over their wages and jobs. However, attention needs to be paid to the development of legitimate control systems in co-operation with staff, in order to enable problems to be identified and sorted out. Staff also need assistance with the process of dispute resolution, otherwise a jungle-law situation can develop!

Bullying can also be a control system in its own right, linked to good stories circulating on the 'grape-vine', and a rule of fear can take hold. Consider the manager in a hospital whose personal control system was to take staff who had not complied, lock them with her in her office, and berate them for an hour at a time. As far as senior management was concerned she was a model manager as she surpassed all the targets all of the time. The control system she used could be justified as a 'proper talking to' but the content was not what one would want. When they became aware of it, the Hospital Board did not condone her control 'system', and she was persuaded to take early retirement.

Structure and the culture web

Johnson and Scholes usefully consider 'structure' separately from 'power'. Usually the structure of the organisation is that which you would see on a diagram indicating how the organisation is made up and who reports to whom. Most organisations have some form of hierarchical structure where the people at the top of the organisation are responsible for those below them. Typically as one progresses up the structure one gets further away from the customers and become more concerned with the internal workings of the group.

Having described several of the elements which make up the culture web, the reader will be appreciating how they intertwine and overlap. When one aligns structure with control systems, we can see how they are often interdependent. For example, in one major city, the Central Library had two managerial vacancies. As a result, one member of staff had no line manager and also no senior line manager directly above them. Because of this gap in the hierarchy, our employee, who felt she might be being bullied, had to make her informal complaint to the Director of Library Services instead of her (non-existent) line manager or her (non-existent) senior line manager. City offices, like many other large organisations, can be quite intimidating places, even to those who work there, and she felt unable to raise the issue informally within the structure in which she found herself. She also did not want to make a formal complaint and so she left the library.

Policies relating to bullying and harassment often seek to use the organisation's hierarchical structure as a mechanism for taking 'informal' action. When many bullies are line managers themselves (certainly in Britain, perhaps less so in other countries such as Scandinavia) the structure may not be helpful. Senior managers will often support their staff – but what happens when a very junior member of staff raises a problem about their line manager who is often also part of the same staff team as the senior manager? Senior managers will, in such circumstances, often defend their junior managers perhaps because their own performance is dependent on their staff, or perhaps because they feel a strong loyalty. Some senior managers work in a culture where they are always expected to defend their immediate staff. If one has a culture where senior managers are always expected to support their junior managers, then a climate is provided where the issue of bullying cannot be raised effectively on an informal basis. Loyalties can be split in these situations. This situation highlights the need for an independent structure to be in place to resolve bullying problems.

In small organisations this is often not possible. The smallness of 'micro' businesses may mean that the owner manager gets very little feedback on their management style – they just keep having problems with staff.

Structure and hierarchy are formal ways of assigning responsibility. It is hard to look at structure without also discussing power, because responsibility can only be discharged when one has commensurate authority and power. Distortions in this balance of responsibility, authority and power, can set up middle managers in no-win situations so that they may resort to bullying. Imagine the young woman who is in charge of one of the restaurants in a pleasure park. She may have tried all ways to bring a sloppy member of staff up to proper performance with no result. She may have tried coaching, extra on-the-job training, as well as other approaches. Suppose she gets to the point where she asks for the person to be removed or transferred (actions perhaps beyond her authority) but her request is refused because replacement labour is unavailable. In such a situation she may well be tempted to use more intimidating tactics with this staff member given that the structure above has failed her.

Power and the culture web

There are many sources of power in organisations. On many occasions within this book the authors have used organisation structure and hierarchy to illustrate how someone can have power over someone else. Position and status might be termed as a formal and legitimate source of power, but these positions can still be abused.

A manager in a local insurance company saw his role as the lynchpin person responsible for the profit margins achieved by the office. In this role he habitually checked all the insurance quotes given by his staff. All new staff were corrected upwards for their quotes, no matter what margin they had achieved with the customer. This meant that all new staff had to get back to their clients and give them the bad news that unfortunately their provisional quote could not be confirmed, and a new deal would have to be struck. Needless to say the manager went through rather a lot of new staff!

At Head Office, the Regional Insurance Sales Manager was perplexed by an unusual situation; the sales figures were reasonable, the margins were terrific, but the turnover of staff was also high. This was unusual as staff were paid commission on sales and high earners typically stayed with the firm. Some commented that the manager must be exceptionally good as he had achieved such good margins with large numbers of trainees. To senior staff, who saw only the profit margins, the chap was a hero. In the local office he was seen as a tyrant and a bully. The office was situated in an area of high unemployment and so many staff did just cope. After the third person left with a nervous breakdown and threatened to sue, the full situation came to light. Here we can see another angle on the need for a balance between responsibility, authority and power. The manager did have the responsibility to achieve results, but he was abusive to staff and he went too far in using his formal power as a manager.

Power can derive from informal sources too. Personal connections to top staff can be believed to insulate individuals from normal systems. Suppose the 'sloppy' member of staff mentioned previously in the pleasure park restaurant was actually the son of the Chief Executive. It is quite possible that the Chief Executive would not want his son treated in a way that was any different to other staff, but those in the middle of the organisation may have assumed (through cultural norms) that the son was to be left alone whatever his performance. This is an example of cultural values literally being invented by staff.

Some people have very strong personalities which they use to threaten others, and this can also act as an informal source of power. Perhaps not too distant from the school-yard bully, these characters are often well known for their sauciness and quickness of wit which can reach intimidating levels in a work situation. Humour can be cruel at times. Frequently this type of abuse of power will be present in peer bullying. However, it is hard for a person to prove the existence of such patterns (as often there is no evidence other than the interaction) and other staff might be unhappy to provide corroborating evidence in case of retribution. Frequently this is labelled a 'personality clash' (Davenport *et al.* 1999). If the organisation does not counsel those involved about the negative effect of their behaviour, condonation through inaction can become evident. If staff leave or complain about a team member's intimidating actions and nothing is done, a cultural belief can develop that bullies can get away with it. If nothing happens, but the bully hears about the complaint, it is likely they will make life worse for the target of bullying who in turn may well become too scared to report it again (UNISON 1997).

Symbols and the culture web

A famous story relating to symbols was that of the new Silicon Valley boss whose first job was to take his door off its hinges and hang it in reception stating 'I'm an open door manager'. Symbols can be extremely fast, effective and hard-hitting ways to influence others and get a message across – think of branding and advertising. What about negative messages though? How helpful is the isolated personnel department that is situated remote from other operational staff? Is their physical remoteness a symbol of their lack of engagement with the staff generally?

Recently we have seen the removal of material with which people decorate their offices such as calendars showing nude models that can be offensive to others. A growing trend is to replace these 'personalising' items with individual notices which are often seen as funny but which have a serious side. For example, the secretary who works for a team of very pressured financial market traders has a notice which reads 'Tell me why it's an emergency after you sorted out the last one on your own!' These notices are symbols for those who display them, give a message to visitors and colleagues, and many of us use them.

We need to be careful about the message we are putting across, however. How confident would you be in the person with whom you want to discuss your experience of bullying who has a cartoon above their desk which reads 'Don't come to me with problems, come with solutions!'? For the target of bullying, who has often already tried all possible solutions, such a symbol may be off-putting. One does need to be very careful on the use of symbols particularly by staff within helping structures – the symbols need to connect with an unambiguous message.

A university promoted their internal 'help' network for bullying (where one could take informal problems) by using stickers on the back of toilet doors in the rest rooms. Over a number of years, this sparked debate. Was it a positive symbol because those who felt bullied could expect the level of privacy from the help network that one achieves in a toilet? Was the university giving the message that bullying belonged in the 'toilet' of the organisation – a very negative symbol? One thing for sure was that people who felt bullied certainly knew where to get the telephone number for an informal discussion. As the university came 'on-line' the stickers were removed and a web site system set up. For some time though, 'Bullying and Harassment' was top of the university directory on the first web-page as it began with a B. Was this a positive message? Was it symbolic that this institution was up-front about this issue and willing to deal with it, or was it first on the list because it was such a problem? This story illustrates the contortions that are possible in this sensitive arena, which in turn reflects how we are thinking about the message through positioning. There is a steep learning curve for the use of symbols that have sufficient gravity and also communicate an appropriate approach.

Putting the web together

In some respects, the culture web works on the principle that the whole is greater than sum of the parts, or $2+2=5$. Behaviours are accepted within a culture and become 'the way things are done around here' because they are supported by more than one element of the web. A spider builds a strong web so that wherever its prey falls in the web, the whole structure works to keep the prey captive – single strands are supported by the others. It is often the same with corporate culture. Anyone working to make change happen knows that single interventions are not enough – instead a wide range of changes (sometimes quite small) working together can achieve a significant shift. A bullying policy on its own will not have a great deal of effect, as our examples have shown. We will be returning to this later in Chapter 10 when we examine the steps an organisation might take against bullying. What is useful about the culture web is that it can be a tool to increase our awareness of how many threads hold in place the acceptance of bullying behaviour.

One culture or many subcultures?

We must also acknowledge that subcultures exist. For example the 'norms' in an accounting department may be very different from those in a sales office. In the latter one might expect a noisier and more bouncy environment. Such different atmospheres will reflect the different types of people one finds in such environments, and the atmosphere may need to be different. What is of concern to us is when the norms of decency and respect for one another vary beyond what is acceptable. How this is judged represents a balance between protecting people's right to free expression on the one hand, and protecting people from a distressing experience on the other.

We must also be sensitive to boundaries. For example the UMIST study revealed considerable bullying in prisons. It is inconceivable that the prison environment does not affect the norms of the staff working there. Boundaries between what is acceptable, where and with whom must be very hard to both identify and maintain in such environments.

Consider the restaurant in which one has a very clear line between the kitchen and the area where customers are eating their meals. The levels of courtesy that are extended to customers may not be the norms that operate behind the kitchen door. Those who analyse culture have to take such differences into account, and may find that extremely strong subcultures need to be dealt with separately. This will be considered when we look at what organisations can do about bullying at work.

Do organisations bully?

A recent and interesting thread of research by Liefooghe has revealed how people see their *organisation* as bullying them, and that although their managers deliver 'negative acts', these managers are seen as just doing their job and reinforcing practices and policies which are originated elsewhere (Liefooghe 2000). Thus the organisation gets blamed, not the individual managers. For example, the strict rules in call centres (where people answer telephone enquires for clients) extend to explaining why new calls are not picked up, or why the operator leaves the terminal at which they are working. All this is monitored by software, and there are supervisors who go around the room and deal with problems. It is a very tightly controlled environment, and some of the operators find it oppressive and humiliating – for example each week they could be allowed only a certain number of minutes away from their terminal to visit the toilet. Liefooghe's results show that staff are almost sympathetic to the supervisor's job.

One could blame the system, but there are other issues too. First, someone would have had to order the software and agreed how supervisors would work with it as a tool. In addition, there would have been a whole set of supervisors and managers who had accepted the situation, and not changed it, or rejected those managers that had made challenges.

With the development of home-working in particular, software needs to be used as a management tool, not as a non-negotiable 'Big Brother'. Software like this is for monitoring workers' activities, but it does not prescribe the supervisors' or managers' actions when deviations are flagged up. Supervisors and managers may need extra help in recognising that they are still in charge and still have responsibility for the well-being of their staff. It may be easy for those in supervisory or management positions to believe the machines all the time.

We should not be surprised at (and perhaps should welcome) reports of adaptation to such potential strait-jackets of personal and professional control. Liefooghe points this out well, showing how some supervisors in the call centres have reasserted the human interface, and interjected laissez-faire into the system. The 'machine' has therefore been tamed and is used as a tool rather than the slave driver.

The authors would assert that organisations don't do things, people do things! We all have the power to question the system in which we work, but elements of the culture web can be felt as powerful deflectors to such challenges, and stories of others who have tried and failed will fuel the concern of those who might consider such action (Denenberg *et al.* 1998). This said, many people experience constraints such as a genuine need to keep a job and fear that sparking a conflict may jeopardise their position

(Davenport *et al.* 1999). These situations emphasise the need for those with security (either personal or professional) to be aware of the problems and take action.

Labelling the paradigm

The central area of the culture web is called the 'paradigm'. Johnson and Scholes (1997) expect us to work around the other elements first to gather data on the culture that we experience, and lastly to get to the point of labelling the culture as a final element. One seeks to uncover what central themes emerge from the various parts of the culture web, and use this in order to label the paradigm. In the call centre example, we have identified elements of over-control that contribute to people feeling abused at work. In other organisations it may be the opposite – a lack of action or passivity. For example, failure to have control systems able to assess the problem allows a pattern of unchecked complaints and staff leaving because of bullying. It will be up to the individual to interpret the culture web diagram and make these connections.

Bullying is usually a symptom of deeper organisational problems, and it is likely that there will be other problems (Rayner 2000b). If one can reach this deep level of analysis, the final effect of tackling bullying at work might reverberate to other problem areas and symptoms.

Chapter 10 addresses actions that managers in organisations can take against bullying at work. What is important now is to acknowledge that organisational norms, practices and processes play their part in sustaining a working environment in which bullying can exist.

6 What causes bullying: the instigators

In the previous chapters we looked at the problem of bullying at work from several different angles. We have tried to give the reader a feeling for the size of the problem, who is involved, and the likely outcomes. We have also identified organisational culture as a factor that may contribute to the presence or absence of negative behaviour and bullying.

Previous chapters have described bullying. We will now seek to analyse the data so that we can start to build up our strategies for dealing with the problem. This chapter begins the process by unravelling the instigators and causes of bullying at work – not an easy task.

Our approach will be to identify possible causes of bullying by examining how the problem may be understood at the following levels: the level of the individual, interaction between perpetrator and target (dyadic interaction), the group, the organisation and finally at the level of society. Such an approach will enable us to focus on a broad spectrum of factors and processes involved in bullying at work (Hoel and Cooper 2001).

Work on bullying in schools will not be used as we see this to be a very different situation, with different power dynamics and different players, to that of bullying in the workplace.

Using examples

We will use examples in order to assist the process of identifying potential instigators for adult bullying at work. As with all our illustrations, changes have been made to protect identities. These two case studies will draw from the experiences of 'Ann' and 'Roy'. Until recently, Ann (44), was a teacher in a medium-sized comprehensive school in the Midlands region of England. Roy (23) is a production worker in a large manufacturing company in the North of England. They provide two different scenarios by which we will demonstrate some of the realities that exist.

In Ann's words:

> Today is exactly two weeks since I went off sick. I can't take it any longer. I went to the doctor who sent me home with a sick-note, certifying that I was suffering from severe stress. Continuous headaches, aching muscles, waking up in the middle of the night in a sweat with no chance of going back to sleep. Life used to be good. I used to love my job, loved watching the kids getting to grips with new knowledge and enjoyed the company of my colleagues.
>
> All that suddenly changed about six months ago when the new head teacher, Martin, took over the school. The last few months have been a sheer hell and it feels like years.
>
> The first weeks he was here, I found him rather charming. Asking how I was and insisting that I shouldn't think twice before asking him for help. Then suddenly he started picking on me in an aggressive manner. Small things became big problems. I gradually realised that Martin had got it in for me. It seems I couldn't do anything right any longer. Everything was criticised, it was like he was *looking* for possible fault with a magnifying glass. I decided I had had enough when he had a go at me in the Common Room. It was so humiliating in front of everyone. I was scared at the time because he seemed quite angry and it was hard to reply.

Roy, about his experience:

> I do like my work and I'm not afraid of putting some extra time in when it is necessary. But I don't feel that my commitment to this place is really appreciated by the company. Despite the fact that I always give at least a hundred per cent, I don't get anything back other than flack from the foreman and the other workers. I guess I am lonely now – it would be nice to have someone to share my lunch break with. There was one, Tom, but he soon had enough and left sometime last year. If they could just leave me alone I think I would feel better. I am sick and tired of their smart comments and stupid games. I have to pretend that it has no impact on me. There are times when I wish I could leave it all.

Instigators at the individual level

Some people seek to explain bullying by looking at personality types. This is especially true for bullies where a cycle of bullying has been suggested, beginning in school, through adolescence and aggressive behaviour at home, and then into workplace bullying (Randall 1996). We should start by emphasising that there is no clear data to support this hypothesis, but, as it is a sufficiently widespread suggestion, it deserves attention.

The idea of such a 'cycle of violence' regarding bullying (Tattum and Tattum 1996) is appealing partly because it is so straightforward. It suggests that bullying behaviour which develops early in life will continue to manifest itself in numerous situations and different spheres of life, of which work is but one (Randall 1996). It is tempting to adopt such an explanatory model of bullying.

Underlying these ideas, though, are various factors that provide grounds for concern. How far is 'personality' and its expression consistent throughout life? Are we unable to change? Many people would claim that they change in attitude and behaviour throughout their lives. Other people see themselves as more or less the same person all their life. Sparse evidence exists for or against the consistency of personality. Many who would suggest that personality is stable throughout life would maintain that we can change our behaviour at any time, thereby separating behaviour from personality (e.g. Hogan and Hogan 1995).

Given that behaviour certainly can change (for example the development of patience or humility), our own experience tells us that thinking of people as stable with regard to bullying behaviour over-simplifies matters. The very fact that many people, irrespective of their role in bullying at school, become successful at work and in private life suggests that we should be careful in drawing firm conclusions (Smith *et al.* 2001). We would also like to warn against the consequences of attaching a permanent negative label such as 'bully' or 'victim' to a child, adolescent or adult and what this may mean for those individuals later in life.

How do we decide on cause and effect?

Let us start with bullying at the individual level and our understanding of the role of attributions. When we try to make sense of events that are important to us we 'attribute' reasons for cause and effect.

Attribution theory (e.g. Baron 1990) suggests that when we have *negative* experiences, we attribute the cause to other people or circumstances. For example, if one has a collision in a car, one might seek to blame the other driver's actions, or perhaps one's own car for mechanical failure – neither of which one would have much control over. In negative experiences we seek to justify and explain our own behaviour by pointing to factors external to ourselves.

On the other hand, we tend to attribute the cause of *positive* experiences to ourselves. For example, if we make a sale, we might think that we have read the situation very well and used a particularly appropriate sales technique. In both situations (positive and negative), attribution theory identifies how we are biased in our explanation of events. Jones and Davies (1965) refer to this skew of judgement as the 'fundamental attribution failure'. This concept of 'attributions' will be referred to in this and later chapters and it will be important for several aspects involved with the analysis of bullying.

It is pertinent at this point to return to the findings of a study that was mentioned in Chapter 4. Baumeister and his colleagues (1990) asked people to recall events where they had been angry themselves or where they had been on the receiving end of anger. The accounts differed dramatically. People who had been angry themselves reported fewer negative feelings and a stronger resolution after the event. People responding to the same question when they were in the role of the target of anger saw the reasons for the anger as external to themselves and had not come to terms with it.

It is not difficult to see how attribution theory may apply to bullying, particularly when we are looking at patterns of what or who is blamed. It is highly likely that the person who is being exposed to bullying will blame their bully or their organisation for events. Equally, the alleged bully may claim that the target is over-sensitive or point out other external reasons for the problem. It is likely to be a negative event for both bully and target, and external reasons are likely to be sought by both parties. The Baumeister study illustrates that this process might be easier for the bullies.

The Norwegian psychologist Kile (1990) spent many years help-ing people get over bullying episodes. He noticed that the longer and more severe the victimisation process, the less likely was the target to admit to any responsibility for events. This opinion also became more set in the minds of targets as time progressed after the event. For the target to survive there may be little choice. Anything giving rise to 'defeatist' thoughts may have to be denied since they may threaten an already faltering self-esteem (Hoel *et al.* 1999). To outsiders, however, the situation might seem very different and investigators often report that the target of bullying was sometimes active (perhaps unwittingly) in escalating the conflict (Ishmael 1999).

In the case of Ann we see how her impression of the head teacher altered from her first meeting with him. Of course it is possible that it was all calculated and planned by the head teacher from their very first encounter, but that is hard to believe. Statements from other colleagues may help throw some light on events. Here are comments from three colleagues who work at Ann's school:

Liz says:

No, doubt Martin (the head) has an authoritarian style, which can be quite off-putting. Whether he is a bully is another matter. Apart from the one time he gave Ann a public dressing down, which I think was out of order irrespective of Ann's previous behaviour, I have never seen him losing his temper. She's pretty sensitive.

Paul says:

Martin (the head) has a huge job to do and I think it's reasonable that he blows off every so often. In my opinion, that's all it is. He had a go at me when my end of year reports were in late. I wouldn't want his job and I think we must be sensitive to his stress. I think Ann over-reacted a bit and took it too personally after he confronted her in the staff room. He shouldn't have said it like that, but essentially he had a point. I think that's been lost in all the furore since.

Darren says:

I think Ann's been very upset by this and she has a right to be. I would really have made trouble for Martin if it had been me in her shoes. So I suppose I think she's behaved very well considering the circumstances. Martin should be pleased he has such a committed member of staff who still soldiers on despite his abuse.

Ann, herself, says:

> I don't like to use the words but in my opinion Martin is definitely a psychopath. There is a rumour that he used to be bully in his previous school as well. I am obviously not the first one. My question is, when will he be stopped?

It can be seen that Ann and Darren are attributing the problem to Martin alone, while other colleagues are including situational factors as well. All investigators of bullying at work will come across many different stories and different perceptions held by people when judging the potential instigators of bullying. We can use attribution theory to understand some of the biases that people have on an individual level.

Bullies are not necessarily psychopaths!

Bullies are quite often called psychopaths or sociopaths. These terms should be reserved for their meaning in clinical psychology where such people seem disconnected to the world in that they act in pure self-interest, lack scruples, and they are apparently without guilt when they damage others. There are very few people like this – perhaps 0.5 per cent of the population, and many are within the prison population (*Daily Mail* 2000).

In the authors' view far too much workplace bullying exists to suggest that these psychopaths, so few in number, are the cause. Psychopaths are very difficult to detect at the hiring stage (Mantell 1994) and will be extremely difficult to deal with (Denenberg and Braverman 1999). Their behaviour is also unlikely to change as they retain their own reality of the world regardless of the other messages coming through to them (which is one basis of the term 'pathology'). This is not to underestimate the awful disruption that a psychopath can cause (Babiak 1995) for those who unwittingly employ them. Undoubtedly psychopaths do exist, but data suggests they are so few in number and so difficult to detect that our attention regarding causes is better spent elsewhere.

Flawed in personality?

We have already made some comment on this, but it is very hard if not impossible to come to any conclusion regarding the impact of personality in general. Bullies are notoriously difficult to study (see

Chapter 4), and our efforts to gather reliable data here will take many years. Many people would see such a search as doomed from the start as the causes of bullying go well beyond individual profiles (e.g. McCarthy 2000).

As regards targets of bullying, some evidence has been produced. An Irish study compared targets of bullying with another group who were not bullied. The second group was 'matched' to the first group by demographic factors such as age, type of job and also home living situation. They were all asked to complete a personality test. It was found that targets were more likely to have a personality profile identified as introverted, conscientious, neurotic and submissive (Coyne *et al.* 2000). This type of personality profile is typical of victims studied in other negative circumstances such as rape, multiple theft and abuse at home.

Whilst this piece of evidence suggests that people with a particular set of personality traits are more likely to report as victims of bullying, it is inconclusive with respect to cause and effect. In other words, whether the traits observed were conducive to bullying in the first place or whether what is being measured is simply the result of the bullying process (see also Quine 1999; Leymann 1996) is completely open to question.

The UMIST study (Hoel and Cooper 2000) also investigated this issue with Professor Peter Smith, an international expert in school bullying. Professor Smith provided questions regarding the experience of bullying at school which were then included in the UMIST survey. The results showed that although it was possible to trace a link, in that people who reported being bullied at work were also more likely to report being bullied at school, this was only the case for *some* adults (Smith *et al.* 2001). However, according to Smith and his colleagues, many victims of bullying early in life are not later victimised at work. Taken together, the results suggest to these authors that factors other than personality and experience in school are important in deciding who gets bullied and who escapes bullying at work. Thus it would appear that the potential for 'screening out' potential targets of bullying, like psychopaths and bullies, is not an option worth pursuing.

Instigators at the dyadic level

Let us return to the story of the teacher, Ann. Liz went on to describe her impression of the relationship:

Liz:

Ann and Martin sort of wound each other up, like clockwork soldiers. You could see how Ann's challenging of Martin about some of the changes he made rubbed him up the wrong way. She wanted to look at every dot and comma, as actually we all did. But I think he got fed up with the time it was all taking. Martin's style seems to be to get things generally right and then make corrections afterward, whereas Ann thinks you should get everything right at the start. Their personalities clash and neither finds it easy to back down because probably they're both right in a way. But Martin is the head teacher and so we have to go along with his approach in the end.

Paul:

I didn't like the way it developed. Both of them are strong people and they both find it hard to back down. That day Martin got at Ann in the staff room wasn't the start of it, they had been irritating each other long before then. The big shame is that they both care passionately about the school and the pupils.

Ann herself:

After he yelled at me in the staff room, I realised he would go to any lengths to get his own way. I don't know why that was such a surprise, because it's obvious when you look at his behaviour. Even in small things he just must get his own way – those sort of people are dangerous, especially in a school.

These excerpts illustrate how Ann and her colleagues are able to see an escalation in the conflict with Martin. This is a fundamental aspect when we look at the dyadic level and is useful because it gives us insight into the bullying process, especially the dynamics. Most often conflicts are resolved at an early stage or disappear. However, if disputes are not resolved they may escalate. As each party gets more involved, it is common for the conflict to start to include issues that are more personal which as we know can be destructive (Jehn 1994). In such a situation, the parties may even forget the original dispute but be 'at war' and seek to beat each other rather than solve the job of work in front of them. This process becomes an escalation of the original dispute which can turn into bullying (Einarsen 1999).

What makes such an escalation happen? The escalation from conflict to victimisation lies as much in dyadic interaction between the people involved as in their personality. Factors such as aggressiveness and self-esteem will play a part here. The social skills of the individuals are crucial, especially those relating to their ability to see a situation from the other's perspective. Such skills will also affect the extent to which the individuals are aware of the effect they may have on others, which is especially difficult in a conflict situation when emotions are running high. For many people, stepping back and calming the situation can be almost impossible when that situation has become a personal vendetta.

Other situational factors will also influence the course of events. In the case of Ann and her head teacher, an imminent visit by schools inspectors may have played a part for both of them in their assessment of the situation. Rather weak exam results for the year before might also have added extra stress to both their perspectives. This would be a good example of two people who were highly committed to their organisation, but who had not paused long enough to sort out their first dispute which then became an escalating pattern, taking on a life of its own.

The situation described in the previous paragraphs suggests that despite the best intent of those involved, events may unfold differently from those desired or anticipated. We can all recall events which turned out worse than we had planned. Most of us have been in situations where we did things we have later regretted.

When people have been asked how they have put a stop to being bullied, it appears that they have been effective at the onset of the negative behaviours (UNISON 2000). This might be effective because it stops the bully getting into a pattern of negative behaviour. It fails to allow the behaviour to be accepted and acceptable. Clearly there is a risk that early interventions could heighten the conflict, rather than end it. We will return to this issue when we deal with what individuals can do if they find themselves in the position of being bullied in Chapter 9.

So far, then, we can see that what happens *between* people may be at least as important, if not more so, than any individual trait or reaction. Our analysis will now progress to the level of the group where such dynamics can be complex.

Instigators at the group level

In the previous section we focused primarily on the dynamics of conflict escalation between two people. However, as we saw in the school example, such dyadic interaction may, at times, be dependent upon other factors such as the response of other group members. Take the example of Roy:

> Peter, a colleague of Roy, explains:

> I remember Roy's first day in the factory well. As the shop-steward I welcomed him to the team. I try to give all newcomers an idea of how things work around here so that they can fit in quicker. Amongst other things I told him the situation on the production bonus. We'd found that if we consistently beat the production bonus then management increased the targets and you ended up doing more work for no more money. So it was important not to break through the production bonus.

> I must admit I was quite taken aback when he rejected my advice by replying, 'not interested' to joining in. All the other guys are affected if one person does this. He also wouldn't join the union, which I suppose fits with his other attitudes.

> At tea break I asked him to join us but he pointed to his watch and replied that it was officially two minutes before tea-time and he wouldn't leave his machine.

> Later that week we discovered Roy had asked the foreman to do something about the men's language, too much swearing you know. At that point we'd had enough. Since then few of us have had any time for him. I know some of the men can't leave him alone and I don't approve of it, but I still think he asked for it, in a way.

From the group's point of view, Roy's behaviour is at odds with the culture and rules of the group. According to Peter, Roy had been made aware of that but was unwilling to fit his behaviour to that of the group. Some readers will find Peter's justification for the group's rejection and isolation of Roy unacceptable. However, that is not the point. As we discussed in Chapter 5, any group or organisation will have their internal customs which they expect newcomers to accept. If these unwritten rules or codes of conduct are broken, rejection from the group is often the result.

Social skills are necessary to detect the norms of a group. If a new manager does not appreciate the prevailing culture in their team or if a new member of staff is not sensitive enough to the signals around them, then this might act as an antecedent for bullying at work. Of course, someone may be aware of the situation, as Roy was, and then choose to go against the trend. We should all be able to be different, as recent work in 'diversity' highlights. Bullying may be one response to a lack of tolerance of diversity.

Some values might be seen by the group as negotiable and others not. For example, one can see how the adherence to productivity rates in Roy's group might have been important. This is an interesting situation, as essentially Roy was doing no wrong by working as hard as he could. But coupled with the tendency for management to raise the productivity threshold once the bonus level had been met, his actions were seen as antagonistic by the group. If management did not have a past history of raising the threshold, then possibly Roy would have fitted in very well.

We should not avoid the point that some people have habits which others find extremely irritating, and as such they can make themselves the targets of bullying at work. Many of us have experienced working with people whose behaviour we found irritating or antisocial. A good example of this was provided by Neil Crawford from his work as a psychotherapist at the Tavistock Clinic in London. A female victim of bullying had been referred to him to talk about being excluded from her group. When he approached the woman's colleagues it turned out that their irritation was caused by her incessant talking which they had tried to deal with on numerous occasions. They had ended up by excluding her as the only effective way of stopping the chatter.

Unfortunately there are many reasons why a member within a group may be seen to not fit in. The example of Roy is quite tangible and we can see both sides of the argument. However, there are other circumstances where issues unconnected to work are cited by targets as the reason they are bullied. Examples such as supporting the wrong football team, having smelly feet, never attending after-work drinks or wearing clothes that are out of fashion have been heard by the authors.

In some situations a bully appears to hound a member of staff, and the group can play a role in this type of situation too. By keeping quiet (perhaps for fear of being the next target) the group can have a role in the dynamics by virtue of doing nothing. The target may feel that the group is against them too (Einarsen 1996), although this may not be the case – the other individuals may be acting out of self-protection. If group members do not take steps to support the target either publicly or privately, it is unsurprising that the target will interpret the group's action as supportive of the bullying, which in a way it is.

So far we have used examples of the group acting in collusion with bullying at work. It must be noted though that sometimes they can help terminate a difficult situation for the target. In 1999, Ford workers at Dagenham in the UK went out on strike to support colleagues who were being bullied. Although some examples at the site constituted racism by supervisors, it was clear from the reports that many white workers were bullied as well as their non-white colleagues. Local trade unionists co-ordinated action, which was in response to groups of individuals objecting to the treatment of others at the site.

What are the factors which make a single group member raise the issue of bullying or join with their colleagues to 'blow the whistle'? Strong personal moral standards are important, as is an awareness that by doing nothing one is colluding with the situation. One should not underestimate the courage of some workers when they do this, as indeed they may be putting their own jobs in jeopardy.

Instigators at the organisational level

It is difficult to discuss the instigators of bullying without bringing in the issue of stress. Many researchers have made use of the stress phenomenon to explain bullying. In particular, German and Austrian researchers have taken this route and consider bullying an extreme variant of social stress (e.g. Zapf *et al.* 1996). The approach taken by the Tavistock Institute (Crawford 1997) views bullying as sometimes one of the products of a 'pressure cooker' organisation. This approach looks for the stresses and strains within the organisation that provide a climate for negative behaviour to be part of a coping process when there is too much pressure. Effectively the organisation is acting understandably within a dysfunctional situation and stress plays a major role.

Before we go any further it seems natural to spend a moment exploring what we mean by stress. Like many other authors, we are using the term stress in its negative connotation as synonymous with distress. In this respect stress is perceived to be negative for the individual.

Although there are a number of models of stress in use, the 'transactional' model of stress has achieved a growing consensus. Considerable work has been done using this model (e.g. Cox *et al.* 2000) and it is helpful to our exploration. This model takes as its focus the interaction between a person and their situation. In order to explain the stress process, attention is given to the role of 'situation appraisal' and 'coping' by the individual. As such the transactional model of stress explains the differences between people as situation appraisal and coping are subjective concepts. This highlights the fact that, when faced with similar situations, individuals will have different experiences.

In the transactional model of stress, 'situation appraisal' refers to how an event is perceived, whether it represents a threat, and to what extent it can be controlled. 'Coping' refers to the way an individual might deal with the situation. Both appraisal and coping are dependent upon the individual's experience and resources available in the situation. According to the transactional model, stress is seen as the *perceived* imbalance between internal and external demands facing individuals, and their *perceived* ability to cope with the situation (Cox *et al.* 2000).

Situational stress

Although the experience of stress is highly individual, in order to understand it, general causes have been sought. These determinants may be divided into three groups, personal, social and situational (Neuman and Baron 1998). We have already given considerable attention to personal factors when we analysed bullying at the individual level. Similarly, in Chapter 5 we explored social factors in the form of cultural influences of bullying. At this stage we will concentrate on situational factors.

Ann's colleague Peter:

> I remember well Ann coming to me last November, complaining about an incident with the new head. I can't recollect the exact details of our conversation, but I remember that my advice was for her to ask for a personal meeting to sort out things before they got out of control. I know it was just before the school

inspection and everybody was tense and exhausted at the time. I remember Ann saying that she wished she could, but she hadn't got the time and didn't think Martin would have any time either.

Several studies of bullying at work have tried to explore what situational factors may bring about bullying. In a German study, it was found that high job demands combined with lack of control over time could function as risk factors from bullying (Zapf *et al.* 1996). These causes are reflected in Ann's case, when there was not a *perceived* opportunity to resolve the conflict at an early stage. The situation, therefore, exacerbated the bullying as it also contributed to preventing resolution.

In Scandinavian studies of bullying, the situational factors that have been most frequently identified with bullying were 'role-conflict', particular styles of leadership, and situations where the workforce has little control over their work (Einarsen *et al.* 1994). These will be examined separately.

The term role-conflict is used to describe situations where individuals are faced with conflicting expectations of the way they do their job. The most obvious example is the situation where there is a demand for an increase in output as well as an improvement in quality. It was found that when such demands were combined with little control over planning and process, then these factors were associated with higher rates of bullying (Vartia 1996). It may be that these factors are central to some bullying – that is, where the bully places extra demands on staff and does not allow them any input into the planning process.

Unsurprisingly, the authoritarian style of management has been associated with situations of bullying (Vartia 1996; Ashforth 1994; Rayner 1999c). An authoritarian style may contribute to the situation, even though it may not directly be interpreted as bullying. For example, it is probably more difficult for staff to alert others to help diffuse a situation in an authoritarian atmosphere. Also blame might be assigned quickly in an authoritarian atmosphere. This might act as a deterrent to anyone raising the issue of bullying as often these situations are ambiguous and complex, requiring more careful examination of events. In addition, discussion between staff and openness of expression are not usually associated with authoritarian styles. Taken together one can see how an authoritarian style might provide a situation in which people feel they cannot raise issues such as bullying at work, as it is too risky.

Combining the factors together

The restaurant kitchen is a good example of how a number of factors may combine to cause bullying (Hoel and Cooper 2000). Anyone visiting a commercial kitchen at the height of production and activity will understand how events can come become very tense. If there are no proper standards for conduct, circumstances will easily degenerate and escalate into conflict.

The typical restaurant kitchen is noisy and hot and often crammed with people, equipment and food. If good-quality food is to be produced within a short space of time there is little room for error in this environment. The fact that many chefs own their restaurants also means that success in terms of profit is an added pressure which can cascade down the kitchen hierarchy.

This situation of time pressure, lack of space, lack of tolerance of failure and the requirement for good quality can add up to provide an environment that precipitates negative behaviour. A British Channel 4 documentary using hidden cameras showed verbal abuse, physical threat, belittling, public humiliation and personal degradation of kitchen staff by the more senior chefs. This programme sparked a debate within an industry where the problem was already well known (e.g. *Guardian* 19 August 1996).

If many knew about such abuse, why had such a situation been left unreported for so many years? It may be impossible for those people inside the situation to challenge such behaviour. If they did it could mean negative job references and no future work. Strong social pressures have meant that many people within the industry have fitted into the industry norms which are themselves abusive.

In most sectors of life, being branded a bully and having your abusive and degrading behaviour transmitted to millions is likely to have major repercussions for ones future career. However, this did not seem to apply to the restaurant industry. Instead, the cook was hailed as an artist whose behaviour needed to be respected with the heading '. . . from enfant crédible to enfant incrédible' (*Guardian* 19 August 1996).

Given such a reaction, will such unpleasant behaviour be repeated by the next generation of chefs and preserve the commercial kitchen as a domain characterised by abusive behaviour? Fortunately, some chefs were appalled by this exposure, and have demonstrated that high-quality food can be produced successfully without the establishment of a dictatorial regime where hurling pans and abuse fill the air (*Hotel and Caterer* 1995). As such, the debate continues.

A pressure situation

Most organisations have the potential to pass stress through their departments as they attempt to deal with pressure.

Phil (Roy's foreman):

> There are times I feel sorry for Roy. I know I have used him, pushed him to do things the others refused to do or speed it up, to increase outputs. But what do you do when the manager made it clear to you that 2,000 units had to be ready for delivery at two o'clock on Thursday come hell or high water. Since Roy is the one most likely, though under protest, to do what you insist on, he will be your target time and time again. And when he gets moving, the others are likely to come into line. Since the other guys don't like him anyway, it's also unlikely that there may be any complaint. In this respect Roy knows where he stands.

In recent years managers have come under pressure for greater accountability. With this a growing number of responsibilities have been transferred or devolved downwards, so fewer people are responsible for more, thus allowing for heightened productivity. In addition, the breadth of responsibility has widened. For example, many line managers have been given the responsibility for tasks which previously were seen as the domain of the personnel or human resources function. Typical examples would be absenteeism control and disciplinary situations.

How many managers have sufficient training to deliver skilful handling in such situations? Such new powers may also lead to a greater chance of unintentionally abusing power (Lewis and Rayner, in preparation).

Greater accountability and de-layering results in changes to career progression. Moving up the organisation is less likely (as there are fewer posts), and the steps are greater for those who are successful. Pressure builds up for those who are ambitious, perhaps similar to those pressures experienced by the chefs who felt that there was no room for error. For many, the opportunity for career advancement has been severely reduced. This means that there is a greater pressure on the individual manager to *continuously* demonstrate ability, with little room for failure. Adding long hours to increased pressure in general, it is easy to see how the situation may translate into abusive and unacceptable behaviour.

With de-layering, career advancement now means that many more skills need to be learned in the new job and few have the perceived time (or perhaps the resources) to go through sufficient training as they might have done in previous decades. The pressure cooker of antecedents for managers is undoubtedly high, increasing the risk that a 'blow-off' is needed.

At the level of society

Programmes such as the Channel 4 documentary on chefs can spark debate in society. Whilst the societal level of analysis is undoubtedly an important factor (Lewis 1999), it will only be covered briefly here as a cause, as this book is about making change at the organisational level. It is possible that societal values, which perform a similar role to corporate culture, are less questioned as we often subconsciously accept them as 'normal'. Therefore, stepping back and providing an analysis is sometimes harder to do than at any other level.

In Chapter 1 we looked at a number of factors arising from globalisation and demands for increased competitiveness, which are likely to produce a work-climate more conducive to negative behaviours and bullying. Individually we are unlikely to alter the persistency of these societal forces, so perhaps we should focus on how we are dealing with them, as that is one thing we can affect.

We must also bear in mind that a large number of managers are identified as bullies and we must pay close attention to those factors which affect them. In Ann's case we saw how some members of staff thought it was important to follow the style of the head teacher in getting most elements for change right, and then correcting on implementation. In other words – the boss is right. British society is hierarchical, and perhaps societal norms to follow the boss's style are more accepted in the UK than in more egalitarian societies such as that of Scandinavia.

The emphasis in some societies on having 'value for money' for products and services has resulted in comparative league tables of achievement. In the UK this is especially prevalent in the public sector where we currently have schools inspections (OFSTED), league tables for local councils (Beacon status), and comparative data on hospitals. With these approaches only a few organisations can be successful and at the top of their league. We need to be careful how we treat those who do not achieve excellence. Have we failed when our local council is not in the top ten for our country? Will there be repercussions inside the council (with accompanying job losses) as a result of being 109th out of 217?

We need to make a distinction between bullying behaviour which is intended to cause harm, and behaviour which aims to 'get the job done' (Crawford 1997). In analysing the role of societal norms, it may be more correct to focus on instrumental behaviour and even instrumental bullying (Hoel *et al.* 1999). We should ask the question 'instrumental to what end?', as sometimes society-imposed edicts hold the key beyond any individual, group or specific organisation. Let us return to statements from the two key players at the school:

Ann explains:

> There has always been disagreement about how we were going about things in our school. Of course, in a staff group of nearly fifty you do not expect everybody to agree. Sometimes disagreements could be quite heated as well, but generally we came to some kind of compromise. However, it isn't that kind of a climate any longer. The rules of engagement have become national inspections and the need to score well on them. If you disagree with any particular policy, you are seen as a trouble-maker and someone who is resistant to achieving success for the school. You certainly can't disagree with the sacred criteria of the national inspection team! Whether a change actually is beneficial to the kids or not doesn't seem to be that important any longer, it's just about scoring points.

Martin's comment (the head teacher):

> I am responsible for the education of over five hundred pupils; to ensure that the educational standards in the school are high and that our targets are met. I have no time for any individual teacher's personal agenda. With the Board of Governors breathing down your neck, there are limits to the level of detail I am willing to discuss with a group of teachers who know little about how to run a school. I am responsible for a good report

from this school when the national inspection gets to us. Their judgement is the bottom line. In my job I have to accept that, and I have to play ball.

We have to measure our organisations' performance, and of course applaud and learn from those who achieve very well. Martin is doing his best to achieve the highest possible marks for his school at inspection. We need to be careful about what happens to the (inevitable) majority that do not appear as 'excellent'. We need to ensure that the results of our measuring system are tools for constructive improvement within our societies, not destructive actions which further undermine sometimes already shaky foundations, and perhaps encourage bullying.

We also need to be careful with some of those values in society that we saw reflected in the response to the programme on chefs. If 'Rambo'-style managers (or chefs, for that matter) are celebrated, one must ask how far that supports the notion that bullying at work is acceptable.

Equally we must be concerned that we do not turn our managers into bullies by providing inadequate training or support as they tackle their ever-changing working environment. Our managers are not super-human, and if it is a societal norm that managers are expected to be so, then we need to do some serious thinking. What permeates all of our comments on this area is our treatment of non-achievement.

Taking this point further, some commentary (e.g. Hoel and Cooper 2001) raises a question regarding the place of bullying within industrial relations dynamics. The authors note that in some American management literature, workers have been referred to as 'dysfunctional' and seen as 'misbehaving' when they have employed tactics to resist managerial control. In contrast, there are rarely any negative labels attached to managers engaged in workplace conflict. According to Hoel and Cooper (2001), these views may contribute to legitimising bullying in the workplace, so that such management literature reflects a further 'institutionalisation' of bullying within our society.

In this section on society, the reader may have noted that we have not made any comment on political activity such as work/social policy setting by government. We have specifically avoided this issue as it is too large an area to cover in a book written primarily for practitioners. Others who are more politically motivated might be able to pick up workplace bullying as an issue and carry it further into the realm of policy setting.

Dealing with complexity

More than anything, this chapter should have demonstrated the complexity of finding causes of the bullying phenomenon. Those who investigate bullying incidents will already appreciate this point. The examples given underline our finding that causes need to be treated with care. If one starts to look for specific causes, one will find them! It is far too easy to jump to conclusions, especially with tunnel vision. It is necessary to gather data on many levels of analysis before judgements are made. Whilst we can more easily influence some causal issues (such as an individual's behaviour) than others (such as societal values), it is important to have an appreciation of all factors.

7 What is bullying?

At this stage, we have presented most of the evidence regarding bullying at work. Almost all the evidence has been accompanied by a caveat: a qualification of some kind which highlights a drawback in method or a restriction to the adopted view. Such caveats reflect the nature of the topic, and reflect the continuing evolution in thinking and practice. This chapter will act as a capstone to previous chapters by discussing directly the problems of how we conceive of bullying. That is, what does the evidence so far mean for the rather simplistic definitions of bullying presented in previous chapters?

The chapter will be structured around the various questions regarding the nature of bullying. Often these questions are simple to ask but rather hard to answer! We will not have all the answers, but we hope that at least we will be able to provide a discursive structure so that readers can formulate their own judgements in a clearer way. We also hope to pose further questions which provide scope for future developments.

Is there one definition for bullying at work?

Generalised definitions of bullying at work were presented in Chapter 1. These were taken from organisations that have demonstrated good practice in the area. A review of existing policy definitions can be found elsewhere (e.g. IDS 1999). Nevertheless, these definitions do not allow for someone, such as a personnel officer, simply to tick a set of boxes in order to determine whether or not someone has been bullied. Such a judgement necessitates the interpretation of definition. The circumstances for bullying can be so diverse (as has been shown already, and as will be highlighted further in this chapter) that we would suggest that these types of definition *are* appropriate for use in policies because they can be interpreted flexibly as situations arise. We began by asking whether there was one definition of bullying at work, and as regards policy statements the answer is 'yes', so long as it is a broad definition which allows interpretation.

Staff in personnel departments need a fairly loose definition which can be applied to many situations. Other professionals, however, may need a much tighter definition because of its potential usage.

The classic examples would be the lawyer who is trying to decide whether to take a case to court or someone conducting a survey who needs to make (usually artificial) definitions so that they can decide how to classify survey responses. It is important to recognise the different needs of these different professionals. While the lawyer will be seeking a definition which demonstrates evidence of bullying behaviours and damage to an individual, someone managing a culture-change project might use more subtle 'indicative' features so that they can judge whether bullying *might* be taking place.

The spectrum of interest could be very wide – from situations where bullying has undoubtedly occurred and harm has been caused, through to situations where one might be suspicious that something is going on. How does one decide where to focus? Some organisations work from the legal point of view and are interested only in detecting and dealing with clear-cut cases. We would see this as a minimalist approach which we would not encourage. Other organisations think they have a moral responsibility beyond the law.

In several countries (e.g. Germany, Scandinavian countries and Britain), health and safety law includes psychological or 'psychosocial' issues as well as physical safety, and as such, bullying would represent a psychosocial hazard. This area of the law requires organisations to show that they have observed 'due diligence' to *prevent* hazards. If one is concerned with prevention, then one needs to have a focus which extends to events *prior* to the bullying episodes.

If we want to extend our definition of bullying to those situations that *could* cause harm, then we will need further discussion regarding what bullying is. We will progress through a more complex examination of the topic by looking at what could constitute bullying at the interpersonal level, at the workgroup level and at the organisational level.

What is 'bullying' at the interpersonal level?

Here we are looking for behaviours experienced by someone which originate from another person or other people. We have seen how researchers who have started at this point have generated lists of negative behaviours and asked whether or not people have experienced those acts in their workplaces. This is perhaps the simplest way of conceiving bullying – where someone reports negative behaviour.

Bullying is subjective

Such a straightforward position means we encounter the subjective nature of bullying since one person may interpret a behaviour as 'intimidating', whilst another person may not. Someone new to catering may find the behaviour of an aggressive chef intimidating. Those who have been in the industry longer may have had more experience of such behaviour and become less affected by it or less aware of it. An alternative source of subjective difference might be found in the role of attributing blame within situations as discussed in the previous chapter.

We should not assume that people's views on bullying at work are static. Over time people may change their judgement for a variety of reasons, usually depending on their experience. For example, we have speculated that targets of bullying behaviour may become 'sensitised' to negative acts so that they become more acutely aware of their presence. One might see this as a highly appropriate defence mechanism to avoid being bullied again. Witnesses may also change their interpretation as they recognise the negative effect of bullying behaviour on others.

Box 7.1

Since last summer the bureaucracy and ever-increasing paperwork were growing at a rapid rate. Hardly a day passed without threatening and demanding faxes from management. I felt totally weighed down by the situation and from enjoying my job I went to feeling like I had never done enough, even if I worked through the night, never taking a lunch break to keep 'on top of things'.

Andrea, telecommunications manager.

It is also possible that people interpret behaviour differently when different people are the source of the negative actions, as has been found in the field of sexual harassment (Jensen and Gutek 1982). We have already drawn on the Tasmanian study (Farrell 1999) which showed that nurses were much more disturbed if they experienced negative behaviour from their nursing colleagues than when doctors were negative toward them.

Such examples show that the recipient is part of the judgement of negative behaviour. We are dealing with a subjective concept where individuals are different and will experience negative behaviours differently. This complicates our understanding of bullying as well as our attempts to improve our knowledge.

The negative reaction

The last study brings us onto whether a negative reaction from the target is required in order to conclude that the person is being bullied. Whilst some work shows us that almost all people find that the negative behaviours we have investigated do 'bother' them (Rayner 1999c), the extent of the negative reaction may be an issue in our definition of bullying.

Whether the reaction of targets should be included in a definition of bullying at work will be a dividing factor for different professionals. A lawyer who is trying for legal purposes to decide whether someone has been bullied will look for damage as a result of bullying, so clearly there must have been a negative reaction. The extent of that damage will be crucial as it will affect the extent of the claim the lawyer might bring on behalf of their client. Someone who is looking for extra ways to improve staff motivation may not be concerned whether people have experienced damage in the same way as a lawyer. They might look for negative reactions that are far more

---

Box 7.2

I know I should have followed my GP's advice and take some days off, but as I knew that the bullying tactic of my boss needed little excuse to oust me from my job and in reality I was afraid to take any sick-leave.

Carole, retail area manager.

nebulous such as not making suggestions for improvement pro-grammes, odd patterns of people leaving the organisation or reduced commitment.

Negative behaviours may not be 'equal'

Unfortunately to date we have been concerned mainly with count-ing the frequencies of behaviours. For example, we know that 'withholding the information to get the job done' is a frequently reported behaviour (Hoel and Cooper 2000). The UMIST study was useful as its large response rate meant that subgroups could be validly examined. The study found pointers that behaviours in some industries were viewed as more serious than in other industries (Hoel and Cooper 2000). This is an exciting finding and supports the notion that not all actions are seen as equally negative. It shows the way to further study that may reveal more of the differences between people and industries which may be helpful for intervention.

Intention of the bully

Whilst we have not allowed the intention of the 'bully' to enter into our original definitions (see Chapter 1), it may play a part in subjective judgement (Field 1996). Our exclusion of intent on the part of the bully provided an artificial barrier because of its implic-ations for an operational definition. We were concerned that if a bully denied they intended to bully someone, then bullying would not be seen to have happened. While we would strongly defend this position for the purposes of definition, intent may hold a level of importance for the players in the situation and our general thoughts about bullying.

Focus groups conducted in several industries (Liefooghe 2000) have revealed that intent is a factor which people entertain when talking about bullying. It is possible that if a 'bully' is perceived as highly vindictive, then the experience of the recipient may be different from that where the 'bully' is seen as just reinforcing corporate requirements.

Intent does need to be treated carefully. In the previous chapter, we examined the role of attribution theory, and this would have a direct impact on the issue of intent as people may use it in their attribution of blame. In addition, the retrospective judgement of intent is a difficult point as we tend to change our perception of events over time.

Labelling

Labelling is perhaps the most contentious issue both for researchers and practitioners. If someone reports experiencing behaviours that would normally be thought of as 'bullying', but does not label themselves as being bullied, how do we treat this? This is common – estimates as high as 50 per cent have been recorded for people who report the experience of behaviours but do not label themselves (Rayner 1999b; Cowie *et al.* 2000).

Perhaps the subjective nature of the experience may account for this, as people do not react in the same way. Equally, it may reflect the bluntness of the questionnaire as a tool which cannot easily reflect the intensity of an experience. It is also possible that some of the people who experience behaviours but do not label themselves as bullied have a lack of awareness (e.g. Adams 1992; Field 1996).

Box 7.3

In an attempt to keep my job, I have worked nearly seven days a week for nearly two years and with more than 1,500 miles driving per month. With hindsight I realise that it was the deliberate tactic of my boss watching me working myself into the ground.

Carole, retail area manager.

Labelling has been an area of importance for those who work with targets of negative behaviour at work. It has been assumed that people who label themselves as bullied are experiencing 'worse' situations than those who do not. Evidence (Hoel and Cooper 2000) shows that this is not necessarily the case, as we demonstrated in Chapter 3. Here we showed how people were affected by their experiences of negative behaviour, regardless of whether they labelled themselves as bullied. This finding is also reflected in large studies on sexual harassment which report the same labelling phenomenon (Magley *et al.* 1999). Their data showed that the experience of sexually harassing behaviours affects and damages people, and we should not get side-tracked on labelling. The UMIST study was the first to provide evidence that we are seeing the same phenomenon in bullying at work, and it is reasonable to see how the two areas should show parallels.

The Norwegian psychologist, Einarsen, urges practitioners not to offer the label to people who they think may be targets of bullying who have not yet labelled themselves as such (Einarsen and Hellesøy 1998). Drawing on many years' experience working with targets of bullying, Einarsen reports that labelling can start a negative process of targets becoming more damaged, as they are more likely to give themselves further labels that are negative, become more 'victimised' and more damaged as a result. Other people (e.g. Adams 1992) consider that labelling the experience is a positive first step to effective resolution of the problem. Clearly, more links into established areas of occupational health would be welcome in order to understand these dynamics further. Research using longitudinal studies and qualitative research on groups of individuals will assist our understanding of the role of labelling in bullying at work.

Persistency

In Chapter 3 we saw how people who are badly affected by bullying have symptoms that are similar to those who suffer from PTSD (which is usually caused by a single traumatic event). Even though bullying is about persistent events, targets can be considered as having PTSD (Leymann 1996; Einarsen 2000).

There may be more links between PTSD and bullying than we have explored to date. Consider the person who has very few bullying incidents to report or perhaps just one. They may have been very frightened by a single experience, or they may be obsessed by it and rehearse it over and over in their minds, so that effectively they experience it several times. Would these people be seen to be bullying themselves? Our notions of persistency certainly do not include such self-repetition or rehearsal and additional work is needed which may provide further insight into the dynamics of how the conflict escalates. One has to remember that targets of bullying have not behaved badly at the start. They may not cope very well, but it would be wrong to pass the responsibility of the bullying to them even if they do obsess over it.

Bullying at the workgroup level

What is happening to other people affects the interpretation we give to our experiences. The workgroup is a 'middle' layer of focus, sandwiched between the dyadic and organisational levels and may affect both. For example, an individual may interpret if someone is bullied by whether others in the workgroup have the same experience. Equally, one might justify negative behaviours at the organisational level if they are widespread and happening in many workgroups. Additionally, by examining the level of the workgroup we are able to broaden the discussion to include local 'norms' of behaviour and introduce context into our discussion.

Context is an important factor in our judgement, as anyone who has changed jobs will appreciate. When we gradually assimilate into a new workgroup, we are inculcated with the norms and expectations of the people in the new environment. As already mentioned in Chapter 5, different workgroups can provide different 'subcultures' in which widely differing forms of behaviour exist. Sometimes these subcultures relate to the type of work undertaken by the team. For example, many sales offices are lively places where energy levels are high, banter flows freely and mutual motivation is typical. They can be compared to accounting offices where the nature of the work usually requires individual concentration and careful attention to detail. In accounting offices one rarely has a background of noise and overt activity as this may distract those working there. These very basic factors will affect the types of behaviour that are enacted, perceived and therefore reported as 'bullying'.

Box 7.4

She said that once again I had been responsible for losing some white-board pens from one of the class-rooms and I reminded her that they were not lost but probably used by one of the other lecturers. When I reminded her about the incident the next week she referred to it as ' just a general bollocking'.

Lee-Anne, lecturer.

Studies of bullying at work vary in their estimate of how many people report being bullied together, but certainly one can expect a range of responses. Perhaps it is a 'worse' experience when one is singled out and bullied rather than bullied with others? If one can see that one is treated differently to others in the workgroup it may help in the labelling of one's situation. This might help explain why people who report being singled out and bullied are usually the fastest to resolve their situation in some way (Rayner 1997). It may be that people who find themselves in such circumstances are able to judge very quickly that something is wrong and therefore take action. Alternatively, they may judge that their situation *is* actually worse than that of their colleagues and for that reason requires more urgent resolution. In short, we do not know.

Being bullied as part of a group is likely to be a different experience from being singled out. If others in one's workgroup experience the same treatment, it may be easier to use that knowledge to take comfort in some way that it may not be 'personal' (Hoel *et al.* 1999). This may have a mitigating effect, making the experience easier to cope with, but we can only speculate on this. However, one could argue that it is worse if the person perceives others having the same experience but apparently being able to cope (even if, in fact, they may not be). In such situations, people may feel worse at appearing (in their own eyes) to be the only person unable to cope.

It is possible that people who are bullied as part of a group are able to share coping strategies. The UMIST study provided evidence that many people, although not all, talk to colleagues about their experience, but we do not know the nature of their interactions. It is possible that these colleagues suggest that they just 'put up with it' rather than attempt to take any action. One should not under-estimate the fear that people have when being bullied, and this may extend to persuading others not to take action in case the situation gets worse. If colleagues do not want any action taken, then it must be very hard to intervene, as one risks rejection by the bully (or bullies) and also by one's colleagues.

Sometimes people report being bullied by several co-workers. Such a context may be very intense in itself (regardless of the behaviours being exhibited) and perhaps may amplify the effect of any actions. Such situations can also be highly dynamic so different people can be the principal bully on different days and also pick on different people. If every so often a target has a 'good' day and avoids much criticism, they may hope that the focus is about to shift to someone else and this may discourage them from taking action.

We have also seen that when colleagues fail to provide support, the target might see these workmates as part of the problem. In these cases, the target might label some people as bullies even though they have not directly done anything negative. Is a lack of action against bullying actually bullying in an indirect form? This might be especially important where one person is singled out from the group and their colleagues keep quiet to protect themselves. How far we should judge the complicity of the colleagues is unclear, but their silence would certainly involve them in the bullying dynamic.

Box 7.5

The head came in, looked around the room and asked the person sitting immediately to my right, would he mind moving so he could sit there. He knew of course that I was going to put forward a different case than his own and I am sure he did that to intimidate me, knowing that I was relatively new to the place.

Oliver, teacher.

Once again, we would only point out the dynamic rather than make any judgements, as the original situation was not caused by these inactive colleagues. There may be many reasons for inaction, and we are hampered in drawing conclusions as most of our information comes from the targets of bullying.

To summarise our exploration of workgroups so far

At the workgroup level, it can be demonstrated that many different circumstances can be labelled as bullying in terms of the players, their number and their level in the hierarchy. In seeking to find an answer as to what bullying is, should we take these into account? Our definitions tend to focus around the target, but the circumstances describe factors outside of the target. If we are looking to understand bullying at work, analysis of factors external to the target such as colleagues and patterns of dynamics may present opportunities.

Some people would encourage us to be very sensitive to the norms of the workgroup and provide solutions that are sensitive to these differences (Crawford 1999). Whilst we would support such a view, should this sway us when deciding what bullying actually is? Is it acceptable that shouting occurs in the restaurant kitchen but not in the presence of the customers? Should different 'norms' for different workgroups be acceptable so that what might be judged as unacceptable in one workgroup could be taken as acceptable in another context? The issue of context has far-reaching implications.

In later chapters we will examine what individuals and organisations can do about bullying. We are at a point in this chapter where there is a strong link with intervention strategies. Do we, in our organisations, decide the behaviours and dynamics that represent bullying and apply them to the whole organisation? Simply put, is shouting always wrong? Or is there an issue of context to be employed which allows for interpretation between departments or sites?

The authors have seen that practitioners, when challenged by unhappy staff, find it very difficult to justify allowing for difference within organisations. It is only after trying to find justification for allowing obvious 'bad' behaviour that many practitioners have reached the conclusion that one should have the same concept of bullying across the whole organisation, regardless of the workgroup, and this is understandable.

It is possible that there is a middle ground, albeit very hard to manage. There may be some behaviours which one decides are always bullying and effectively 'bans', and others to which one may take a looser attitude. Different work groups will be sensitive to different dynamics and if one can monitor effectively, one might be able to steer a course where such differences can be used. This discussion brings us to our final level of analysis regarding bullying – that of the organisation.

Bullying at the organisational level

At the end of Chapter 5 we considered whether or not organisations can play the role of the 'bully'. In that chapter we examined how some systems seem unfair to employees and suggested that managers or supervisors may be excused as bullies by their staff because they are, indeed, reinforcing polices. In that chapter we took the stance that systems and policies can always be tracked through to people – the designers of the systems, those who agreed to them and finally those who implicitly agree with them by their enforcement. Such a 'systems' approach (where processes are the main focus of examination) can be very helpful in exposing a long string of decisions that together can add up to a tough working environment for staff. For people involved in the decision chain, it may be very difficult to speak out as they perceive themselves acting against others, which may be seen as a risky move.

We are steadily increasing our knowledge about how the perception of unfairness at work can de-motivate employees. Unfairness is about comparisons between oneself and people who work for the same employer and also perhaps those who work for other employers. At the organisational level, apparent unfairness may appear from many sources. For example, it may be connected with policies that are not uniformly interpreted or certain individual managers who decide to create unique working environments.

This returns us to our discussion of how far one should encourage diverse environments within the same organisation, and it is a difficult point. For a multinational corporation, the cross-cultural data would suggest that different cultures value different aspects of the working environment. In 'low Power Distance' countries (Hofstede 1980), employee participation in decision making may be highly valued, whereas in an environment which expects bosses to be different (high Power Distance) then consulting with staff may actually undermine a manager. Such arguments may justify difference.

For those who employ people within the same national culture, finding justification for different treatment within the organisation can be harder. Is it contentious to encourage a participative management style in one area of France and not another? It may be acceptable, but it will need to be thought through as it may have to be explained and justified at a later date.

When we examine the organisational level it is necessary to include corporate objectives and other influences that the decision makers are having to balance in order to achieve a well-run firm or public organisation. All employers are concerned with costs, and commercial enterprises also need to consider profits and where those profits go. Public sector entities will want to demonstrate good value for their customers in terms of the funding they receive and private enterprises will focus on profits (even at a minimal level) as well as costs.

Employers have to balance their responsibilities to their various stakeholders which include staff, customers, suppliers and perhaps shareholders. These responsibilities could be seen to compete with each other at times, and those in charge must find a pathway through. Is bullying acceptable in times when the organisation is fighting for survival? Whilst we cannot ignore such challenges, accepting them would be the start of a slippery slope and could be abused by those looking for an excuse for bullying regimes at work. Interest in this topic will reignite debate that has existed for many years (it forms the tenets of modern socialism and Marxism, for example), is always worth re-examining, and may affect our definition of bullying at the corporate level.

In discussing what bullying is, we must return our focus back to the more tangible aspects within the organisation. The reader should be aware, however, that the questions raised in the previous paragraph may haunt him/her, as these questions are central to this topic.

Bullying cultures

What is a bullying culture? Is it one where many people experience the behaviours outlined in Chapter 3 or Chapter 5? Or are the very high incidence organisations (such as the prison service in the UK) not really bullying because the norms are so entrenched that people have stopped complaining and accept that they work in a culture that is 'tough'? If people re-label their environment in this way, does that mean it is not bullying?

If a whole group of people appears proud of being able to survive a 'tough' working environment, should we, knowing the damage it can cause, consider taking any action? Or rather should we see it as their working choice and leave them to cope? Should we be bothered that people are recruited into 'tough' environments at all? The authors would suggest that we do need to examine such corporate matters for reasons which come both from a legal 'duty of care' position and also a moral point of view.

Further levels of analysis

The last comments bring us to further levels of analysis: those of the national (e.g. Lewis 1999) and international (e.g. Archer 1999). This text will not enter into these debates since more specialised texts can be used by the reader to allow for a local interpretation at the reader's convenience (e.g. Brown 1998). As one progresses upwards to such broader levels, one can expect changes in expectation and norms to be slower as more people are involved. What is considered bullying at a workgroup level, therefore, is likely to change faster than at a corporate level, and so on. When very large sections of populations are taken, such as national norms, change can be extremely slow!

Summary

In this chapter we have tried to get to grips with the subjective nature of bullying as well as the rather artificial parameters that have been used by researchers to describe the topic. This is a new area of enquiry. A book written in ten years' time may debate the dynamics and explanations that have been offered rather than establishing data indicating that we take this topic seriously. Unfortunately we are not at such a point yet! However, the nature of that which we call bullying will, we suspect, remain debatable for many years to come.

8   Who is to blame?

The contentious title for this chapter has been chosen because these words are important for many people involved with this topic. It is understandable that we seek to assign blame when waste and damage occur. Often we do this at the point where waste and damage have occurred already, and, for a variety of reasons, we look back to assign blame. The purpose of assigning blame might be to locate the blame away from ourselves, or it might be that while we are not involved ourselves we might seek to learn from the situation in order to prevent it happening again. Alternatively we may want to fire someone! Whatever the reason, the apportioning of blame can be an emotional process, and we should be aware of this from the start.

There is a more constructive side to apportioning 'blame' – if one can find the cause, then one can act appropriately. But 'blame' is a word that is loaded with negative connotations of judgement. Different roles carry different responsibilities. Blame and responsibility are bound together and logically people get blamed when they have not met their responsibilities. In bullying at work we have noticed many circumstances when events do not treat people quite so fairly and blame is assigned without real cause. This is an emotive topic, and blaming is one of its side products.

When we examine the responsibilities various parties have in the process we will adopt the same structure as the previous chapter, namely examination of the interpersonal level, the workgroup level and also at the level of the organisation. In these analyses we must stress that it is individual action that matters, and that our own exposure to this topic has led us to a realisation that often people look to others to be responsible for action in combating bullying at work. Quite often everyone is looking for other people to take action, rather than doing it themselves, which can result in getting nothing done at all. When we move away from the individual level, we will continue to stress the importance of people in their roles within the organisation.

At the individual level

Of course, we are all responsible for our actions. If everyone completed their responsibilities at the individual level, bullying would not exist as the persistent set of behaviours which we see causing damage today. In the last chapter we unpicked the complexity of bullying and we saw that the pressures for the acceptance of bullying can come from many places. The factors that are at work extend beyond the individual to a much wider set of parameters including the workgroup (and its approval), the corporate norms that are espoused at the organisation level, and the society which reflects national norms. All these factors can impinge on the individual and affect their behaviour.

Thus, while we are discussing individuals, we should remember that their lives are not lived in a hermetically sealed bubble! Individuals act in context; that is, they work with others and within a power structure with rewards and punishments of different kinds around them – some of which they value, and some of which they do not. This section of the chapter outlines the fundamental responsibilities that we have to ourselves and each other in the context of bullying at work.

Box 8.1

I never took any time off work even though I felt sick nearly every morning just thinking of the day ahead of me. Despite the unhappiness of the situation, the experience has made me stronger convincing myself that I am able to take on a more demanding job even though my previous manager time and time again made it clear that I would never go any further.

Elizabeth, teacher.

Making sure we are not the bully

All of us have boundaries beyond which we are not willing to go. Taking an extreme example, most people would not hit a colleague at work. Physical violence is not the norm in most workplaces, perhaps because people take on board their responsibility to control their physical actions in this instance and it is not acceptable. Extending the example further, some people would not join in if their workgroup was belittling a colleague, ever. The socialisation process that forms our decisions regarding the boundaries of our behaviour is complex and not for study here, but we must acknowledge that boundaries exist for us all and we are all different in the boundaries that we keep. It would be wonderful if our boundaries included always treating everyone with dignity and respect regardless of the circumstances, but as we have seen, it is impossible to prescribe such behaviour.

The authors would suggest that we are all responsible for knowing the effect our actions have on others. Such knowledge may occur after the event. Unless you are an extremely conservative person in your actions, it is likely that you have acted inappropriately at some time in the past. Most of us are able to pick up the cues from other people and, perhaps with a little embarrassment, apologise, attempt to explain and resolve not to do this again.

Let us analyse the previous sentence which could be considered an example of 'good citizenship' behaviour. We are responsible for being aware of our actions. Part of that is opening our eyes and becoming sensitised to feedback from others. This is a step that some 'bullies' fail to achieve – they permit themselves not to check out the effect of their actions and therefore never complete the feedback loop that indicates that their actions could be causing harm. We should all do this, and many would say that we are responsible for this as a matter of ordinary social action.

The second step, having realised that one's actions are causing offence or are inappropriate in another way, is simply to apologise to those around us, admit the mistake and clear the way for another set of actions. This is hard to do in a work environment for some people. If one works in a 'blame' culture or a 'tough' culture, for example, such admissions of guilt may change one's personal standing within the group. Sometimes such admissions, viewed by the individual or their work colleagues, are seen to reflect a form of weakness or to be a fundamentally flawed act. In a 'blame' culture the group will not let you forget your transgressions. In a 'tough' culture the individual may be extremely geared to never failing or being seen to fail.

The final step is that of changing our behaviour. This can take reflection and effort if the change is approached from a 'learning' perspective. It might mean that someone has to analyse themselves in order to find the reasons for their behaviour and deal with them. Often people cannot or will not put energy into such action as it involves considerable thought and self-critique. Other people may simply 'ban' themselves from ever acting that way again, not giving any attention to underlying motives or the resolution of any conflict within themselves.

For the employer it is probably unimportant what approach any bullying employee takes, just as long as the behaviour stops. The more self-reflective approach may cause additional problems to the person in the short term, but may present a solid long-term solution and perhaps a change of attitude as well as behaviour. The behavioural approach may cause the individual to wrestle with the stress of denying themselves the behaviour they want to exhibit (for example getting very angry and shouting at people), and such tension may take its toll as a stressor in their performance.

The process described previously assumes that the individual notices that their negative behaviour toward another person or group has caused distress and that this is not what they intended. Unfortunately there are people who do seek to cause distress. Sometimes people want to 'stress' others rather than 'distress' them, and indeed the dividing line is slim. This might be the case for managers who perhaps have considerable stress passed on to them, and who may feel better – when everyone around them shares their own tension.

In Britain there is a game called 'Pass the Parcel' where children sit in a circle and hand round a well-wrapped package. Each time the music stops, a child unwraps a layer and feigns expectation that they might find a present, but usually they do not. Pass the Parcel is a game of positive expectation almost always followed by mock disappointment, as all that is revealed is another layer of wrapping. What has been described previously might be seen as an adult game at work that could be called 'pass the stress parcel'. Together, the group of children eventually unwrap the present. Spreading stress around at work, though, rarely has a final reward, except that the project, the emergency job, or the current panic comes to an end.

Good managers will stress their staff, but only to the point where those staff can perform to their best abilities. In the same way, a colleague who has difficulties at home may bring their problems into work. Too much pressure, as highlighted in Chapter 3, causes *distress* and has negative consequences. We all need to contain our

---

Box 8.2

As it is, I have no social life and rely entirely on my partner for all domestic matters. She has had to take on weekend work in the local supermarket in order to be able to pay the bills. We have also had to sell the car and may have to move to smaller flat, as the income doesn't pay the mortgage.

Derek, ex-NHS.

---

stress and our own emotion so that we can manage our relationships with others effectively. Sometimes called 'emotional management' (Goleman 1996), this is a new area of academic enquiry which shows great promise, although it is too early to suggest as an easy intervention strategy since too little is known about the concept.

To summarise, at the individual level we are certainly responsible for checking out how our behaviour affects others. We then need to act on any negative feedback and make changes. This may not be as simple as it sounds because of the pressures within the environment that might prevent us from apologising or choosing a different behavioural strategy.

Responsibilities for the targets of bullying

If we are the *targets* of negative behaviour, do we have any responsibilities? It could be said that we are responsible for informing the aggressor that we have been hurt or affected negatively in some way. Once again, this can be very difficult in reality. If one thinks of what it is like to be frightened as a child, then this is the experience of some people who are being bullied. As children, many of us found that the safest road to take was to keep quiet in such circumstances, and indeed this is what some people do as adults. If one's self-esteem is rather low and one is being bullied, then too much courage could be required to tell the bully or bullies that this is unacceptable. We would not want to blame any victims for failure to take action, because they have not been responsible for starting the process. They are coping in ways that they know best, although such ways may not contribute to solving the situation.

As targets we are responsible for trying to protect ourselves. The authors would encourage every target of bullying at work to tell someone in the organisation what is happening. This should be someone who can do something for them, or perhaps a colleague who might know who can help.

We have seen that many targets leave their jobs as a result of their experience. For many this will be the best protection that they can see – to remove themselves from the situation. This will be easier for someone whose skills and situation allows them to get another job quite quickly or if they are fortunate enough to be able to survive financially without working. If someone does leave the situation, we would strongly urge them to take responsibility for letting the organisation know the reasons for leaving. This can be in writing or in a more informal way, such as a telephone call. Sometimes targets assume that management know more than they do, and at least this action will make the target feel that they have done as much as possible from outside the organisation.

Box 8.3

It all happened after a rather trivial discussion over lunch and a male colleague followed after me in the men's toilet hurling abuses. A third person already there told him to leave me alone. When the abuse continued I answered back with the result that he pushed me up against the wall. If it wasn't for the other colleague I am convinced he would have hit me. When I went to personnel to file a complaint I was told that an investigation would be undertaken under the condition that I saw occupational health. When I returned to work next day I was talked to as if I was mentally retarded and I was told that the investigation had been inconclusive.

Paul, fire service.

Responsibilities for witnesses of bullying

As individuals who may witness or hear about bullying at work, what are our responsibilities? Sometimes targets see silence on behalf of witnesses as complicity in the bullying process and can blame them for being part of the problem. Similarly, blame can be attached to people whom the targets choose to tell about their situation and who appear to do nothing to help the targets. For others who have knowledge of the situation, there is a responsibility to bring the problem to the attention of someone who has the power to do something about it. There are many different suggestions in the next chapter for what one might do, but we would suggest that you are responsible for informing the target of your actions.

We would suggest that witnesses and others who are not involved directly have no responsibility to solve the situation, and that they should not suggest to the target that they will do so, otherwise expectations may be raised. The problem lies with those who are involved directly and the managers and professionals within whose roles such actions are included. Witnesses may contribute, but assuming the role of mediator or 'saviour' to the target is not a responsibility.

At the group level

It can be very hard for targets to confront a whole group of people and tell them their behaviour is unwelcome and damaging. Even if several people are being bullied by several other people, standing up to a group is extremely difficult, and it is naïve to expect someone to carry this through.

At the group level, there will always be witnesses. The group which is responsible for the negative behaviour may be encouraged by a lack of action on the part of witnesses or those who know about the situation. They may see silence as acceptance or permission to carry on. In such a circumstance it is imperative that witnesses understand and undertake their responsibility to let others know what is going on, partly because it will be especially difficult for targets to do so.

At the level of the organisation

In Chapter 10 there are many suggestions for actions that an organisation can take. Most organisations will comply with the law, but many see their responsibilities as extending beyond the law. We would suggest that this chapter is read carefully by anyone who holds any organisational responsibility, as blame will be assigned by staff, even if your organisation never gets taken to court.

The organisation should have a policy that relates to bullying at work, disseminate it, be able to demonstrably implement it, monitor it, audit the process and act on any feedback received. All of this should be done with senior management involvement.

This description follows a fairly process-oriented pattern, since this is how the law operates. Once again we need to remind readers that staff can be highly judgmental and may be assigning blame. Incidents of bullying are marvellous news for the gossips in the organisation. If staff get to know about the situation, the story may take on a life of its own. Employers should expect that staff will assign blame, and in such circumstances the organisation may be put in a very difficult position as staff may make judgements on information which is incomplete, with the 'full picture' unable to be revealed on grounds of protecting confidentiality.

The original promises on confidentiality need to be kept by the organisation. In some circumstances, the stories that circulate about the incidents can be very damaging to people. Indeed it might appear instead that it would be less damaging to review the confidential nature of some information and be quite open about what has happened. This could be done with all parties consenting, and it is usually done only in extreme circumstances. The principle used here would be that of duty of care to protect employees from harm. However, if the initiative to release information is refused, then the confidentiality must continue.

Summary

People often seek to assign blame in workplace bullying situations. Many people will have opinions, and the authors' experience is that people are quite happy to voice their judgements. Sometimes the process of dealing with the blame is as hard as dealing with the bullying incident itself.

There are lines of responsibility that can usefully be used to protect ourselves against blame. We can all take responsibility for checking that we are not bullying others by being open to feedback, acknowledging mistakes and amending our behaviour accordingly. Individuals who are involved professionally will need to be clear on where their responsibilities lie.

Targets should tell someone, but we would *not* see targets as responsible for solving the situation in which they have found themselves. Instead, we would encourage targets to talk to others so that they can get assistance with solving the situation. Our principal duty of care is to ourselves, and where no help is at hand, employees may think that leaving the organisation is the only solution. If this does happen, then we would encourage targets to feel a final responsibility toward those left behind in the organisation and write to let a senior manager know of the circumstances of leaving. There are ideas for what individuals and witnesses can do in Chapter 9.

At the organisational level, managers should not underestimate the role of blame within the situation – grapevines of communication exist within most organisations but staff must be encouraged to reach their own conclusions. The organisation has some responsibilities in law, and we would encourage a proper policy and procedure to be adopted as this is a way of demonstrating fairness and equity which will protect the organisation and its employees.

Our next two chapters focus in detail on what can be done. Chapter 9 will look at actions open to individuals and Chapter 10 will examine steps that organisations can take to combat bullying at work.

9 What can individuals do?

This chapter is addressed to all readers. Some of you may be directly involved in a bullying dynamic as a target or a witness. There may also be some readers who have been accused of bullying. Many readers will be professionals who are being asked to deal with the problem, for example as a human resources specialist or as an occupational health nurse.

Most readers will have some experience of negative behaviours at work, directly or indirectly. For most of us there will have been times when we have asked ourselves whether we treated someone unfairly or inappropriately. Very few of us will be able to claim with certainty that our actions at work have never left others feeling upset! This chapter should be a good opportunity for self-reflection and revisiting some of our own decisions or the actions we have taken in relation to negative behaviour at work.

In previous chapters, we have shown that no industry or occupation is 'bully-proof', and also that situations vary. This chapter intends to provide some useful advice, and should be used selectively, depending on your own circumstance and experience.

We have divided the chapter into three sections. The first addresses those who are the targets of bullying, and this is followed by a section for those who are witnesses. We then turn our attention to people who are concerned that their actions may be bullying or who have been accused of being a bully. We finish with ideas for all of us that may improve our workplaces.

Deciding whether or not you are being bullied

Some readers will have picked up this book knowing that they are a target of bullying, others may still not be sure if they are experiencing bullying at work. Some people may feel only 'slightly' bullied, and others badly abused. It is not easy to come to terms with being a target, and can take a considerable time. People vary in this process, some keep their experience and feelings to themselves, too embarrassed to let other people know or unable to find the words. Others people react quite differently, involving others by constantly talking about it, but they may still be unsure how to label their experience.

A common experience of being bullied is a feeling of confusion that goes hand in hand with a series of strong and often conflicting emotions such as anger and sadness (Adams 1992; Randall 1997). This emotional turmoil is well described in several self-help books aimed at targets of bullying (e.g. Adams 1992; Field 1996; Namie and Namie 1999; Marais and Herman 1997). In Chapter 3 we outlined the trauma that can be identified with bullying and how it may manifest itself emotionally as well as affecting health and behaviour. For some targets of bullying, these reactions represent a collapse of their fundamental assumptions about the world and themselves.

Guilt and self-blame are often part and parcel of this experience. Wilkie, an Australian researcher and bullying activist, accurately refers to this process as 'internalisation of abuse' (Wilkie 1996). Unfortunately this can also become part of the dynamic if the bully exploits the target's feeling of guilt and self-blame and uses these emotions as a means of further control (Field 1996).

We also showed in Chapter 2 that the actions taken by targets to tackle bullying may seem to be futile, and so one needs to be very careful in giving advice to someone who thinks they may be bullied. Care needs to be taken, not only with what one suggests, but also with making sure that one does not raise a target's expectations that the actions will solve the problem.

The second UNISON study (2000) provided some suggestions from those who had stopped bullying. The theme that came through in these reports was that confronting the bully may be a positive move if done early in the process when the bullying has not become an established part of the working relationship. Thus the message for those who are uncomfortable with some recent changes that have made them feel as if they are being attacked is to address it with the perpetrators, as this might be a positive move. When approaching the 'bully' targets will need to be clear in their message and very level in the way they deliver it. It will be critical for the target to keep calm and retain their focus in this encounter, and not lose self-control, which may add power to the bully's position. It is also important for targets to be realistic in their expectations – the bully may not respond immediately, but they should indicate when they will be able to.

It may be pertinent here to remind readers that bullying can be seen as a process of escalating conflict. It is fundamental that everyone avoids escalating the conflict further. In particular, it is important not to get into games of attack and counter-attack. This is such a dangerous path to follow and we would urge anyone to desist as it is at the heart of the bullying process. It is therefore crucial not to hit back by, for example, pointing out their weaknesses, starting rumours about them or in any other way adding fuel to the bully's fire. This is why we have stressed that a 'level' approach needs to be taken by targets, even at an early stage, so that further escalation can be avoided.

If you are being bullied

So, what are the options for someone who is in a situation where the bullying is already an established pattern? We would suggest two things before any decisions are taken. The first is to expect that the bullying will continue. It is probably a stable situation and will need action if it is to change. The second issue is for the target to think through what they want out of the situation.

For many targets, their primary concern will be to see their life return to 'normal'. For example they may want to keep their job and for the bullying to stop so that they can get on with their life. Other people may think that getting an apology is important. Others may want a public admission of wrong-doing by the bully or bullies. For some, getting vengeance is paramount and they would like to see their attackers suffer in some way. As one proceeds through these different outcomes, it is clear that each is harder than the last to achieve and is likely to take more time. We would encourage anyone thinking of retribution to think twice, as they might be accused of bullying!

We would recommend that targets carefully weigh up what they want. None of the previous ideas will be ideal and all will carry a cost. The calculation should include an awareness of the cost to one's private life as well. If there are people at home that will be affected, then perhaps they should be part of the discussion and decisions, as their support (or lack of it) might be central.

Doing nothing

There may be times when ignoring someone's hurtful remarks or negative behaviour may be wise, particularly if it is a one-off event, for example if the perpetrator acts out of character due to personal circumstances which you are aware of.

However, if the behaviour is persistent, ignoring it is unlikely to make it go away. Targets should look at the costs of doing nothing, which may include being gradually worn down psychologically, which in turn can affect health and relationships. Looking through other courses of action in this chapter and balancing the costs with the benefits may provide other options.

If the behaviour is established we would not normally recommend doing nothing because we would be concerned about the long-term damage to the target. We also have another concern, which is that, if you keep silent, the organisation may remain unaware of the problem, making it more likely that the bully will be unchecked, and use the same behaviour with others as well.

Confronting the bully

Should targets challenge the perpetrator's conduct by confronting them directly? If the bullying is recent and a dynamic between you has not yet been established, there are good reasons to believe that bringing it to the attention of the perpetrator(s) may make it stop. Stating the effect of the behaviour and its unacceptability will, in some cases, be enough to cause the perpetrator to change. It is quite possible that the offender may not realise the effects of their actions, and a clear signal from the target may be enough for the behaviour to cease with or without an apology.

It is also possible that *quickly* signalling this with a small and businesslike remonstration may send a message to the bully to move off to 'easier' targets. We would hope that anyone who achieves this becomes sensitised to such tactics and helps others who might also be approached (see 'Witnessing bullying', later).

If the idea of confronting the bully or bullies strikes fear into the target, then it is possible that this fear is well placed. It may be wise to identify what that fear is based around – for example could the bully fire them, or might they make things worse, and in what way? Working though the target's strengths, weaknesses, opportunities and threats can be an extremely enlightening experience. Many self-help books on harassment provide frameworks for this, and we would encourage their use. Also please note that some books marketed to women are excellent and apply just as well to men (e.g. NiCarthy *et al.* 1993).

Another facet to consider is the level of support that might be given (or not) to the target by others. Colleagues may be unwilling to give support because they are too afraid or vulnerable in other ways. If the bully or bullies are colleagues, then thought should be given about the level of support available from the boss.

When the bully is a manager, targets often report that the bully receives support from the senior manager and this can present a real problem. Remember that support for the bully may originate from respect as well as from fear. So, if there is a likelihood of such support we would advise against a direct confrontation. Our data has shown that confronting the bully is a risky business, with a large proportion of targets becoming further isolated and losing their job as a result (UNISON 1997). By contrast, if the target feels strong enough to stand up to the bully and is likely to be supported then a direct confrontation may be the best move.

There is a middle ground, which is to confront the bully or bullies with someone else. For example, if the bullies are colleagues then support can be sought from the manager, followed by a meeting where problems with the bullies are aired while the manager is present. If a boss is the bully, then support may need to come from more formal channels which are considered later.

Using silence to combat the bully

One way of not escalating the conflict is simply to be silent. This is a calculated silence rather than 'doing nothing' (Field 1996). It is reminiscent of Gandhi's passive protests where silence was used to try to undermine the dynamic. This may be appropriate for some people in some circumstances, but it will not be for everyone. Silence can irritate and escalate the conflict, especially if it is to do with something that is genuinely needed for work. Also silence needs to be used with the right person or people, as one does not want to involve people outside the dispute, who may think such behaviour extremely odd, and possibly disruptive. Be careful not to attract valid criticism.

Using informal methods to get help

A golden rule for safe intervention is to do nothing that escalates the conflict further. This can be extremely hard to achieve, but there can be opportunities to intervene by using informal processes. In the next chapter we encourage organisations to develop informal methods for people to use if they feel bullied at work. Anecdotal data suggests that few targets want to go through the process of making formal complaints (Field 1996; Ishmael 1999), and would prefer to use informal methods to change their situation.

Going to the boss

Informal complaints systems usually involve an employee going to their boss in the first instance. Unfortunately, in Britain most bullies are reported as managers, so this cannot be used, but there will be some people for whom it can work such as those who feel they are being bullied by colleagues. If the manager will take the time to listen, and is sympathetic that the patterns of small events can accumulate to cause targets distress in bullying situations, then this can be an excellent path to follow. The second UNISON study amongst civilian workers in the police force found reasonable rates of resolution when managers were told about bullying situations (UNISON 2000). To a large extent, use of a boss will depend on the target's confidence that they will be listened to.

Going to personnel

Informal systems may extend to sounding out personnel or human resources staff within the organisation. It would seem there are few risks of the situation getting worse if this method is taken (UNISON 1997, 2000). If advice on what to do is needed, then personnel may be willing to help. Ideally they could have 'a quiet word' with the people who are causing distress, and not even divulge this action to the target.

Our data suggests that targets rarely get to know what, if anything, happens as a result of visiting centralised staff such as those in personnel or human resources (UNISON 1997, 2000). The authors suspect that a great deal more action occurs than such professionals are willing to reveal to targets. Unfortunately, by not telling targets about any action, we also suspect that the target's feelings of inadequacy and isolation are reinforced (Rayner 1999a). One must acknowledge that personnel are in a difficult situation because they cannot admit any organisational liability for what is occurring and so they tend to be rather silent in this process. If it is of any comfort to targets, they should remember that much more may be going on behind the scenes than they get to hear about.

The most important point about going to personnel is that the target has let the organisation know their predicament. By doing this, the target has effectively passed on a duty of care to those staff, who then should take some action. Once again we must not raise expectations, because the feedback that we get from targets is that 'nothing' happens as a result of such visits. We would still strongly encourage targets to go to personnel, but reckon that they will not get the feedback they want.

Going to the union or staff association

Many organisations have a trade union or a staff association. Understandably, most British trade unions will not act for staff regarding events that have happened prior to them joining the union. However, most representatives will have an informal chat and help in what way they can. Their knowledge of dealing with similar situations might be very useful to hear.

In particular, we would suggest that targets discuss their expectations of what they want to happen as a result of their actions. This is a form of reality-testing and some targets who seek a full-scale apology for their treatment may get a realistic estimate of its likelihood. In addition, as representatives have had experience of dealing with the organisation, they can provide helpful tips for the best channels to follow.

Using occupational health

Many larger organisations have occupational health staff, although their presence in smaller organisations is sparse. But if they are accessible, their use is encouraged. One should remember that these staff are often geared to dealing with physical health problems. However, in recent years, the effects of stress have received more attention and such staff can usually listen to and give some excellent advice to make life as manageable as possible. We have little data on the effectiveness of occupational health practitioners to help with interventions.

This said, these staff might use their own network of internal contacts to alert personnel or other staff that there is a potential problem, and so more action may be taken than the target is aware of. We also know that the issue of bullying is currently receiving considerable attention within occupational health circles. It is, therefore, hoped that such staff may gradually play an important role in bullying intervention, for example risk-assessment and data-collection on bullying. However, for such a potential to be fulfilled, occupational health practitioners would need to clarify their relationship with the employer in order to act more independently (Hoel and Cooper 2000a).

Using volunteer networks

Finally, your organisation may have a scheme of 'contact advisers' who provide an independent ear and act as sign-posters for options within the organisation. Always check out the level of confidentiality these people can provide before you go into significant detail with them, as they may be not be able to give you full guarantees. However, they are usually very helpful and certainly one loses nothing by visiting them and seeing what they can offer. They will have experience with other staff in the organisation with similar problems so their advice may be invaluable to hear and we would encourage a visit.

Making a group complaint

A substantial number of people are bullied together with others in their workgroup. In such instances, it may be tempting to write a joint letter of complaint or to set in motion some form of collective action against the bully. Such moves may be tempting but have often proved unsuccessful, with threats of dismissal the result. If your group is bullied we suggest that you bring it to the attention of personnel, the local union or staff organisation. Any steps you decide to take can be clarified with them. It may be that your situation is not unique and that the union already has plans for raising the issue themselves.

Going outside the organisation

Some people see their local doctors to get advice. Once again, while we acknowledge that the doctor's main expertise is in physical conditions, these professionals can be very helpful in highlighting how important it is for targets to solve the problem before serious effects manifest themselves.

Doctors may suggest counselling or other remedial work, and of course their advice is usually sound and worth taking. We would point out that these professionals will be keen for you to take some action to resolve the situation otherwise they will be helping someone to cope who could become more ill as the source of distress continues.

Some caution is indicated as, unfortunately, the bullying literature has examples of targets who report not being believed by their doctors. The obsessive behaviour combined with absence of physical symptoms have misled doctors with no specialised knowledge of bullying to diagnose mental illness or personality disorder (Lennane 1996; Leymann 1996). Instead of connecting the individual's condition with their work situation. Recent years have seen a much higher priority being given in the UK by GPs, psychologists and occupational health doctors to bullying, within their professional conferences and updating programmes. Raising awareness is to be welcomed, as these groups will be less likely to misread the symptoms of bullying as a result.

Organisations which are dedicated to helping people involved in bullying situations have been set up, and we have listed some of these in Appendix 3. For example, the Andrea Adams Trust helps all parties including targets, witnesses and employers who are seeking resolution to problems. A useful development is on-line access, so a fair amount of advice can be obtained from abroad too.

As well as seeking help regarding the bullying from outside the organisation, targets should think about keeping strong on a personal basis. According to Davenport *et al.* (1999), a way of strengthening well-being is to meet and engage with people in activities which are likely to boost self-esteem. Self-help books suggest that keeping fit is a good idea – almost working from the outside in – and we would support this as an idea.

Using formal methods to get help

The decision on whether to use the formal complaints system is difficult to take for many targets. If the formal path is followed, the situation certainly will change as now there is more at stake, including people's jobs.

At this point many people decide to leave the organisation, as they think that they cannot cope with the formal complaints process. Once again, targets should be encouraged to take advice.

This might relate to the potential success of their case, as well as to the personal costs of pursuing it. For many targets, the effect on their partners and family will play an important role and they may decide to leave the organisation rather than place further strain on home relationships.

Quite often personnel and trade union staff say that they despair because so few employees are willing to take the formal complaints route. One can sympathise, as personnel and unions may have tried but failed to help informally and may be keen to use an official complaint so that they can deal with the alleged bully or bullies properly. Naturally, the authors have great sympathy with such frustrations. It is with reluctance that we would suggest that targets should weigh up this decision to move to formal measures carefully, because of the extra stress that can be caused.

Some people feel that they have invested too much in the job and the organisation simply to let go and leave without taking action. Others may feel that they are failing themselves, their colleagues, as well as the future of the organisation if they do not see the process through to a formal complaint. This can be exacerbated if the target knows that they are not the first or the only person to have been bullied. If someone reaches this point, then it is rare that any choice is ideal – there will be a cost and a benefit in all options.

Taking notes

As one gets into formal processes the issue of evidence takes a high profile. Formal procedures do require evidence. Many self-help books suggest that anyone bringing a formal complaint should take notes on the behaviours they encounter and collect any other evidence of the situations which cause distress as soon as they feel badly treated. An extreme variant of evidence collection is advocated by Field (1996) who suggests taping conversations with the bully or bullies, as otherwise they are likely to deny any accusations. We would encourage all evidence gatherers to be aware of the legalities of their actions since they will not want to attract criticism by collecting evidence illegally.

Getting evidence does present a problem for most targets of bullying. The authors worry that if people are not going to use the evidence that they collect (such as progressing to a formal disciplinary hearing) then its collection may actually add to the distress. For some people it may be very helpful to write out what has happened to them (e.g. Adams 1992; Field 1996) but for others it may have the opposite effect. Some people feel they re-live situations as they write them down, so rather than being a cathartic and helpful process, note-taking can add to the experience of negativity. We are also aware of people who have found that the process of collecting data has become an obsession in itself, and, as a result, they have felt completely dominated by the bullying in all areas of life.

Leaving the organisation

It is possible that there is little to gain in pursuing a complaint through formal or informal routes. Perhaps the costs and risks far outweigh the possible gains, so leaving the job seems the best option. Of course, such a decision will also be dependent on the likelihood of getting a new job, which will include factors such as the local labour market.

Targets will be concerned with getting a good reference. Sometimes it may be unwise or impractical to leave before alternative employment is secured. Other times, targets may feel that their situation is deteriorating and it is better to leave with a good reference before it gets worse.

It is pertinent to remind readers that many people who respond to surveys on bullying at work also report being bullied in a previous job. We can conclude that targets of bullying do find work elsewhere. Many people leave because they want to put the incident behind them. This is a healthy choice as leaving the organisation may be a conscious move to something better.

Going to the law

Without doubt this is a major decision. It will mean considerable commitment on the target's part, financially and emotionally, with many disappointments. Most people who have taken this route report that going through British courts can be as tough an experience as the bullying itself. That said, most people who have had their day in court apparently feel better for it (Witheridge 1998). Financial settlements are rare, but do occur (the 'successunlimited' web-site carries an up-to-date coverage of legal cases).

The major help-lines for bullying (such as the Andrea Adams Trust) carry excellent material for people thinking of legal action. This is a specialised field and we would encourage you to be guided by them in your choice of lawyer. Good organisations will not directly recommend one lawyer, but will provide a short-list of specialised people.

Getting back to 'normal'

Most people who have been bullied report that the incident has changed their lives in some way. Therefore, 'normal' is probably not what it was before the bullying occurred. However, there is life after being bullied.

We are aware of several people who have become so tied up in getting retribution or court settlements that they seem to have spent a very large number of years unable to free themselves of bullying at work, well after they left employment. If you are worried about this for yourself or anyone you know – do be concerned, as it is a real danger. Kile, a Norwegian psychologist, asked his clients 'Is seeing justice done worth 20 years of your life?' (Kile 1990).

Witnessing bullying

For the person who witnesses someone else being bullied a number of options are possible. Let us explore them systematically.

Validate the experience

We have emphasised that many targets of bullying may find it difficult to define what is happening to them, and may be having self-doubts. In such situations it is vital that witnesses validate the target's experience. Whilst the target's explanation of events may not be flawless, they can be supported in their main argument that they are being treated in an unacceptable manner.

Assess risk

If one is willing to talk to a target about their experience, they are likely to ask for advice. We would encourage a systematic appraisal of their options, and guidance to further advice when the answer is unclear. One has to respect, however, that in the end their decisions are their own.

Witnesses may quite correctly, feel that they run the risk of becoming targeted themselves. It may be tempting to 'lie low' and pursue a strategy that is strictly non-interventionist. This is particularly true when the bully is powerful, well known for his or her bullying behaviour, and likely to be supported by the organisation. By avoiding drawing attention to themselves, witnesses may think that they can stay 'out of the heat' and avoid becoming the next target.

Unfortunately, such a strategy is likely to back-fire in the long run. Each time a bully gets away with their behaviour this must reinforce the notion that such behaviour is acceptable. As such, non-intervention by witnesses can act as an encouragement to the persistence of bullying.

Even in situations where there is little room for manoeuvre, it is important that witnesses privately tell the target about their awareness of the situation and give support and empathy on a personal level.

Protecting oneself

Supporters will be of little help to the target if they also become a target. Conscious or unconscious attempts by targets to draw others into the conflict must be resisted. Providing bystander help will be invaluable, but we would advise against getting involved beyond your role within the organisation.

It is worth emphasising that giving support to a target of bullying can be emotionally draining. They are often distressed, and can ask for support at times that are not altogether convenient. Clear boundaries should be set – perhaps whether conversations outside working hours are acceptable. It is probably better to provide a small amount of consistent support than be an occasional avenging angel!

Be prepared for a bumpy ride, and remember that the target's experience will be worse than that of the supporter. Reviewing the victimisation process and how disabling it can be (see Chapter 3) could be a worthwhile task.

Involving others

If a supporter decides to go to personnel either with or on behalf of the target, it is important that the effects of the behaviour on the target are presented as the focus. We would also suggest that judgements about the bully's behaviour are not made, for example by suggesting that they are a psychopath. Such comments might bring the judgement of the supporter into question, which will not help the target.

Similar approaches to unions or the other internal staff outlined in the 'target' section of this chapter could be considered. The issue of bullying on a more general level could be raised, thereby possibly contributing to a process where knowledge of and attitudes about the problem may change within the organisation, hence reducing bullying.

Assisting targets with formal procedures

If the target decides to file a grievance against the bully, they are likely to be in need of practical help. Advice may be very useful in a situation where emotions are running high. Helping the target put forward a systematic and logical grievance which will be taken seriously by the organisation may be very useful. Assisting the target in any formal hearing about the case may also be invaluable. Again it is important to make sure that it is the target and not the supporter who is the focus, otherwise the supporter may run the risk of being labelled a trouble-maker.

Witnessing your own group bully

A substantial number of people are bullied by their peers. It can take a lot of courage to challenge the behaviour of peers, particularly when the bullying is conducted by more than one person and there is a chance of being ostracised. However, at times, some bullies may have convinced themselves that what they are doing is 'only a joke' or 'just a bit of fun'. Being gently challenged by a group member may be enough to stop the behaviour.

Much more difficult are those situations where the target could be seen to have provoked a reaction by not fitting in with group norms, like Roy in Chapter 6. Here two jobs need doing. First, the issue has to be raised with those bullying, emphasising that, whilst the conduct of the target can be criticised, the bullying behaviour is unacceptable and has to cease. Second, the target needs to be challenged that whilst the bullying behaviour was unacceptable, it was in reaction to the target's own behaviour, and some middle ground needs to be found for everyone to compromise and work peacefully together. This can be a tough argument to make, and we will return to it in our final chapter. How far should the rights of the individual surpass that of the group? We would hope that there is a compromise to be found, and certainly one should be sought.

Who should undertake such interventions? We would suggest that if there is a manager or supervisor, then a witness could be best served by getting the manager to intervene. There is always a danger that asking the manager to deal with the situation may spark resentment from the team that the witness has been 'telling tales' and that they think the group should deal with it themselves. If the witness thinks there might be a danger of this, then they would be advised to work without the manager, as if the team were self-managed. If the team is self-managed, we would suggest the witness pulls in other neutral team members who can bring the parties together. If none of these seems realistic, then a visit by workgroup members to personnel, trade union or other helpful source could be in order.

Whistle-blowing

When every attempt to raise the alarm within the organisation has failed, more drastic methods such as going to the Board, the shareholders or the media might be considered. We would strongly suggest that advice is taken from anti-bullying organisations (see Appendix 3) on such moves before any serious plans are made. The risks involved in such actions are very great, and certainly extend to losing one's job.

Actions such as going to the press may be interpreted negatively by the employer: they may become defensive and counter-attack. No one wins in such battles (except, perhaps, the media!) and the situation can become very nasty. It may be terrific to be famous for a day, but hard to get another job afterwards.

If you are accused of bullying

As the word 'bullying' enters the general vocabulary of more and more people, it is likely that it will be used inappropriately, sometimes even in a joking manner ('come on, stop bullying me'). It is, therefore, important to make a distinction between those incidents where someone is playful and those situations where the label is used in a serious manner.

Take it seriously

We would suggest that you take the accusation seriously. The first task is to listen carefully to what is being suggested. Accusing someone of bullying is a major step for most people and you should assume that something is wrong. Whilst this type of accusation is likely to alert your defensive mechanisms, we would suggest that you should avoid responding to the accusations there and then. Certainly resist any temptation to flair up, threaten or make loud counter-accusations!

A more low-key but positive approach would be to take notes so that you can reflect when you are on your own. This will have been an emotional session and you are unlikely to remember it accurately without written notes. If you think the situation demands an immediate response we suggest that, in many cases, being drawn into a heated discussion would only make things worse. Remember that your accuser will have had time to think this through, and that if you have listened to them carefully, a suggestion that you also need some time to think things through should be acceptable.

If possible, we would encourage the presence of a silent witness who can be another set of eyes and ears for you. The ideal person is someone who can be trusted to provide honest feedback about the situation – not necessarily someone who thinks you are wonderful! They would also be a good 'honest friend' to be able to discuss things with afterward. If you can do this, the person who is bringing the accusation should be told that a silent witness will be present.

Afterwards, try to consider the accusations rationally. You may be surprised by what has offended people, and it is possible that you have been inadvertently misunderstood. If that is the case, a simple apology and a resolution to change your behaviour to other norms is likely to be the most straightforward course.

You may, of course, be totally outraged that someone has been offended by what you might see as completely innocuous incidents. Be careful. Remember that someone is in distress. You are not the judge and jury of when it is alright for someone to be offended, or upset. Remember also that we are looking at behaviours and their effect, not intent. Intent is important to people when they are trying to resolve incidents, and emotions may be defused when it is known that there was no intent on your part to upset anyone. However, a change in behaviour will be expected. An ideal starting point is to ask those who have put forward the accusation to suggest what behaviour they want instead.

There is also untold benefit in apologising to people. Most bullying involves people who need to work together, so at little cost, relationships may be repaired. The fact that you have picked up this book suggests that you are open to change.

When the accusation is false

Bullying is tough to prove, but it is also tough to disprove. The authors have good contact with trade union representatives (who represent those accused of bullying as well as targets), and they have real concerns for people who are falsely accused of bullying at work. If the accusation of bullying is considered to be false, the line manager should be informed immediately. Silence is dangerous. It is unlikely that the problem will disappear of its own accord, and if the accusation is not dealt with, it is quite possible the problem will escalate. If the 'target' decides to bring a grievance, the accused will be better placed if openness has been pursued from the start.

What we can all do

Not all readers will face a problem of high levels of negative behaviours in their working lives. However, throughout this book we have argued that the current work climate and the high rate of organisational change is likely to act as an antecedent of negative behaviour. We would further suggest that no organisation is 'bully-proof'. We all need to keep the issue of bullying alive even when no obvious problem exists, and we would be well advised to work through prevention strategies at an individual level.

Before we move to the next chapter, which has the organisation as its focus, we would like to finish by suggesting some preventive routines for all of us to engage in as individuals.

Monitor ourselves

In the previous chapter we discussed the issue of blame and responsibility. We argued that all of us are responsible for our own behaviour regardless of the circumstance. Perhaps this is a good time for self-reflection. Can you respond positively to the following questions?

- Are you aware of how you come across to your staff, colleagues and boss?
- Do you ask for feedback on the way you behave?
- Do you pay attention to your own emotions whilst at work?
- Is your body-language in tune with what you are saying?
- Do you join in when jokes are made at someone else's expense?

Whilst these questions are applicable to all of us, they are particularly important for anyone in a position of power or authority.

Challenge negative behaviour in others

Small comments that are intimidating, derisory or might give offence need to be challenged. You do not need to do this in an aggressive way. Humour can be a very gentle way of nicely alerting someone that their behaviour is not really acceptable. Most people would avoid wanting to be known as a PC (Politically Correct) bore, but the occasional challenge, in ordinary ways, in ordinary work circumstances can help redefine what is acceptable behaviour at work.

Becoming an activist

We can all volunteer to participate when personnel or other group initiates programmes concerned with bullying at work. By lending support, keeping discussion going in the office, and contributing to awareness in whatever small way we can, together we may improve the climate in which we work.

10 What can organisations do?

We are still learning about how to deal with bullying at work. Some ideas in this chapter are fairly obvious, such as the need for a policy on bullying; other ideas are more controversial and we will highlight the arguments both for and against their implementation. Within this chapter, you will also find ideas which may be more effective in some organisations than in others, depending upon the industry, the culture and the particular site. We trust that readers will use their discretion in applying the suggestions.

Whatever action practitioners may take when dealing with bullying, it is crucial that they monitor the effect of their actions. Research in this area has revealed some findings that have surprised those conducting the studies. We have learnt that we must never assume or take things for granted. Too often our logic can be reversed and completely different conclusions or explanations reached.

In addition, the levels of fear that bullying at work can generate in targets mean that sometimes people do not react as one might predict initially. Bearing these factors in mind, it is of utmost importance that monitoring is undertaken at site level so that, at the very least, one can check that things are not getting worse. This chapter includes a section on monitoring and offers a variety of possible parameters that might be considered by the practitioner.

This chapter is structured so that the more procedural issues are dealt with first. These will include policies relating to bullying at work, and the informal and formal complaints systems that might be put in place.

We will then progress to the more 'fuzzy' areas where we will examine how one can create a culture in which to minimise bullying at work. It is likely that readers will have different circumstances in their workplaces, so we have split this section into three. The first section is concerned with the very difficult situation where bullying at work is seen by staff as endemic and a complete turnaround is required. The second situation is where 'pockets' of bullying are spread around the organisation. The final circumstance is where the levels of bullying are low, and the main concern is prevention.

Policy and procedures

Policy and procedures are the starting point for organisations that wish to install fair and equitable practices (Merchant 1997). For those of you who find this an unexciting prospect, please read this section, because you don't have to take a dull and mechanistic approach! It *is* possible to generate a bureaucratic strait-jacket for staff to follow, but such systems are often counter-productive. What one seeks is a system that works and is helpful to participants. In this section we will highlight the basic steps as well as providing some tips and traps for applying good practice in the area of bullying at work.

For those of you who are seeking to protect your organisation from legal redress, we will be following a variety of sources that make suggestions on compliance systems. This includes 'best practice' advice from British organisations such as UNISON, NASUWT, European commentary, and input from the USA such as the US Federal Sentencing Guidelines for Organisations (Chapter 8). Of course there is considerable overlap in advice from bodies involved in helping organisations to develop policies and practice. As many readers will be concerned that their organisation conforms to inter-national standards, a wide range of parameters have been utilised.

Generating the policy: what it should contain

A policy is a statement of the behaviour and/or principles that one seeks to encourage or discourage. Examples are provided in Appendix 2, and an excellent review containing numerous examples can be found in the IDS Report *Harassment Policies* (1999). Having a statement is important, as it provides a document that is highly tangible and to which everyone can refer. The statement should include:

- a description of the behaviour that is required (e.g. supportive, protecting of dignity, etc.);
- a description of bullying behaviours (with examples that cover overt types such as intimidation as well as the more subtle such as systematically being set up to fail);

- a clear statement that the organisation will not tolerate bullying;
- what consequences could be incurred if someone is found to be a bully (e.g. it is a dismissal offence);
- referral to a procedure or other system which indicates how bullying will be dealt with;
- identification of a senior person of good standing who is ultimately responsible for the policy, the procedure and their implementation;
- a referral point for any queries regarding the policy itself;
- details of how the policy will be monitored and audited.

Any reader who has been involved in generating policies will know that the previous suggestions on policy are identical to those for racial or sexual harassment, for example. In the drive to avoid shelves bursting with policies and procedures, many organisations do not have a policy statement exclusively geared toward bullying. Some organisations have a generalised 'harassment' policy which covers issues to do with race and gender as well as the more generalised psychological harassment that is bullying.

Other organisations have found that the 'list' of negative issues that they want to include in a policy relating to negative behaviour at work has become very long. Such lists might extend to ageism, sexual orientation, specific religious issues, the receiving and giving of bribes, unfounded whistle-blowing, etc. Some organisations take a different approach and, rather than label all negative behaviour, they focus on behaviour they seek, such as honesty, integrity and the protection of dignity for all. These Dignity At Work policies stress the positive behaviour that everyone can expect to receive and is expected to give to others. They have changed the policy from one which provides a list of 'don'ts' to one which provides a list of 'dos'. It is too early to comment on whether Dignity policies are more effective, but one can see their attraction from a streamlining position.

We can say that this is an area for judgement by those working the system. Most organisations have tried at least two of the approaches, and we would not want to suggest any single formula for policies. Most organisations find that just inserting 'and bullying' into a policy that exists for racial or sexual harassment does not work. To be effective they need either to have a separate policy or to rewrite a wider policy to include bullying.

We suspect that the choice of policy is less important than its implementation and this will vary between organisations. However, if your policy does not work, there are several other approaches you can try to re-examine your strategy.

Generating the policy: who and how

The wording of policy statements can be difficult, as one seeks to avoid ambiguity whilst at the same time providing room for the many different forms that bullying at work can take. Although it may take longer, we would suggest that as many people as possible help to generate the policy statement. It is important to involve representatives from any group within the organisation which might be involved with the policy. For example, if there are trade unions or staff associations representing staff, they may be involved in handling complaints, and their inclusion in devising the policy and the procedures is fundamental. Health and safety representatives may receive enquiries, as may occupational health staff. Think carefully about who might get involved with implementation of the policy as the greater the level of understanding and input, the better the system will work when it is launched.

We would argue that the inclusion of a wide group of regular staff (from different levels as well as different functional areas) can be an enormous benefit as it starts a debate about the nature of (un)acceptable behaviour in one's working community. Fuelling such a debate will allow issues to surface early (that will surface later regardless) and highlight areas in need of further attention. This process will allow discussion and also time for as many people as possible to align themselves with the policy as well as 'owning' the policy as 'theirs' rather than seeing it as a missive from on high.

Someone senior should be included in the team for policy formulation and implementation. In many countries, someone senior *must* take final responsibility for activities related to legal areas associated with bullying at work (such as health and safety and also discrimination). It is prudent to ask such individuals to co-ordinate activity from the start so that they can *demonstrate* that responsibility has been taken at a senior level. As the policy has a champion at a senior level, not only is it taken more seriously by all staff, but it has a guardian at the most senior level, such as at board meetings.

Disseminating the policy

Once this policy has been generated, it needs to be disseminated properly. If a wide range of people have been included in the composition of the policy, you will already be aware of areas of concern which will need special clarity for interpretation. For example, managers may want to be clear on the boundary between 'firm and fair management' and 'bullying'. Other staff may seek highly prescriptive guidelines such as a definitive list of bullying behaviours that is actually unrealistic in a work setting.

You should think through who staff might go to for an explanation of the policy. Having consistent replies to common questions is crucial from the start, otherwise a lack of commitment will be inferred and there is a danger that staff will judge these efforts as tokenism. Hopefully, such people will have been involved in the preparation of the policy in the first place and will just need a short further session agreeing verbal guidelines. Remember, however, that there will be a point at which the policy has to be interpreted and staff will not be able to provide answers that are set in stone.

Some organisations disseminate by simply circulating a policy on paper. There are dangers in this approach as there is no feedback as to whether people have read the policy or understood it in the way that it is intended to be interpreted. It is much safer to allow for interpersonal contact through launch events or training sessions where there are staff representatives present, for example from the trade union. If the policy can be embedded into the day-to-day business of the organisation, so much the better. For example, if there is an operational cascade communication system, a session could be used to discuss the policy. Dissemination events should not be voluntary as these will tend to attract the enthusiasts and miss exactly those people who need to attend. Some system of ensuring that everyone has attended needs to be in place. Those who join the organisation at a later date can have the policy included as part of a formal or an informal induction programme.

There is a real danger that policies languish and become ignored and forgotten unless some profile is maintained. Disseminating results of monitoring processes can play a major role in reminding people of the policy's existence. We would suggest such reminders should be contrived at least every six months.

Procedures

When developing a policy, the accompanying procedures will also need to be decided. It is likely that a policy will necessitate the training of staff involved in the procedures (such as investigators or mediators), and that this will need to be thought through so that implementation can be completed quickly.

The key: informal procedures

As we learn to deal effectively with bullying at work, the role of informal processes has emerged as a key issue. Initially most targets of bullying do not have any thought of taking their organisation to court. Instead, there are consistent reports that targets simply want to keep their jobs and want the bullying to stop. We also have some evidence from people who have succeeded in stopping bullying. A characteristic of these reports is that actions have been taken *early* in the bullying process (UNISON 2000). When the conflict escalates it may be harder to stop, and targets' desires may shift toward retribution. It is therefore in everyone's interests to stop the bullying as early as possible.

When dealing with bullying, low-key and low-level interventions that are timely and effective are needed to prevent an escalation of the situation. By identifying problems before distress is caused (and the conflict escalates beyond a retrievable point), harm and disruption to the business can be minimised.

The most informal level is that of good management and care on behalf of all staff. Are we all concerned and proactive if someone is worried by another's behaviour toward them? In the last chapter we examined what individuals can do, and this needs to be brought forward to the organisational level in order to ensure such individual actions are supported and encouraged. Are all staff listened to seriously when they raise issues concerned with bullying at work? Do managers know how to respond if such concerns are voiced to them? Do managers have the skills to raise the issue of bullying in a low-key but effective way without contributing to the conflict escalation? In order to ensure this training may be necessary and would be worthwhile.

Any procedure that outlines informal routes must make suggestions that go beyond raising the issue with one's manager, as often the manager is seen as the bully. One might suggest raising it with the higher levels of management, but evidence shows this too can be difficult (UNISON 1997). It may be 'safer' for staff to raise the issue with a neutral person who has been assigned this responsibility, or perhaps someone in personnel or another aspect of staff support such as occupational health or staff contact scheme members for example.

Whoever is selected as the first contact point for informal enquiries should have the ability to take further action if appropriate. Such action might mean approaching the manager of the alleged bully, to set in motion low-key interventions which bring the issue to the attention of the alleged bully in order to discover their side of the story so that a balanced picture might be achieved. These interventions do not need to be full-scale investigations. Instead, with the aid of people with sufficient social skills and knowledge of those involved, a way through the conflict may be found so that the policy is adhered to as part of the ordinary day-to-day business of the organisation.

Some organisations have networks of staff in place who undertake to be 'contact advisers' or 'buddies' for staff who want to discuss their situation and resolve questions regarding bullying. Typically, such networks draw from volunteers and need careful managing in themselves, issues to which we shall return. Very often, however, their role is not specifically to intervene in any way but to be a positive listening ear for the person who may feel that they are a target of bullying. In this way they differ from the people mentioned previously who can set up low-key interventions informally.

Originally, contact schemes began (in the UK) as support networks for those who thought that they might be targets of harassment. They vary considerably, but best practice points to several important factors. First, the group should be drawn from volunteers who are screened by interview and then trained. Unfortunately, this role can appeal to the gossip mongers within the organisation, and those approaching contact advisers need to be assured that their information will be treated with respect.

Confidentiality is an issue that applies to all strategies for dealing with bullying at work, and it is certainly one that can affect the contact adviser's role. Often, the role of the contact adviser is that of a 'friendly ear' with whom the person can discuss their situation. The contact adviser can point out (but not be proactive and give advice) the various routes available to the person and discuss what action they may take. In these discussions, the contact adviser may hear details of incidents that are examples of clearly unacceptable behaviour by colleagues or managers. How far can they keep such details confidential? What happens when many people complain

about the same person within the organisation and report distress
that really should be getting dealt with at a senior level?

According to British law, anyone who becomes aware of something
that is a danger to the physical or psychological safety of a fellow
employee has a 'duty of care' to report such information to someone
who has the authority to deal with it. For contact advisers this may
mean that they cannot respect confidentiality, but rather have a legal
duty of care to pass on the information they receive if that
information shows that harm is being done or could be being done.

Employers currently avoid this by two means. First, within the
contact scheme no one reveals the name of the people who are the
source of the negative actions, hence there is no clear information
on identity. Second, some employers use an outside agency to run
the contact scheme, as 'outsiders' do not have this duty of care to
reveal the sources of potential harm within the organisation. Both
methods can favour the organisation which really does not want
to deal with bullying. Those in power *can* insulate themselves
from knowledge which alerts them to difficulties within their
organisation and, if they do not have the knowledge, how can
they act? Contact schemes can be 'elastoplast' systems that
provide marginal help for those in difficulty but contribute little to
resolving a problem.

Several organisations who do want to deal with bullying have
given up trying to get around these points. Instead, they make it
clear to those approaching contact advisers that the advisers are
obliged to take action. Thus they restrict the scheme only to those
staff who are willing to take their complaint further. Clearly there
are positive and negative elements to all these approaches and, once
again, often it is how the scheme is run rather than the rules on
paper that make it useful or otherwise for staff. We would suggest
that organisations which use outside agencies, perhaps as part of an
Employee Assistance Programme (EAP) should request full feedback
so that the problem is not just put in a holding position, which will
probably lead to further escalation and potential damage to those
involved.

For contact schemes to be effective, they need to be linked into
the other mechanisms within the organisation so that they are a
dynamically useful measure for preventing bullying. If they are not,
targets of bullying can become more frustrated as they fail to
resolve their situations and the contact advisers themselves can
carry a great deal of stress if they try to act as emotional buffers.

One problem associated with all stages of supporting targets of bullying is that initially targets are often reluctant to complain. We have seen how this worry has solid foundations as situations can worsen as a result of making a complaint. So, in what way can the 'helpers' progress the situation to a fast and positive solution when the target is unwilling to complain?

The solution is to evolve a discreet but effective informal system of handling such information. Here, the 'duty of care' can act as a helpful vehicle (as it was meant to) by placing a burden on those who hear about negative actions to make their knowledge known to others. The law is a good reason to raise and resolve situations through ordinary good management. We should emphasise that such informal systems need a light touch, otherwise they can exacerbate the bullying in the way feared by the targets – by providing irritation to the bully and escalating the conflict further. Although the complaint may emerge from contact schemes (safety advisers, trade union representatives or occupational health advisers) probably the line managers will be at the centre of dealing with such informal complaints.

Ideally all staff need to be able to raise the issue with the alleged perpetrators in a way that is low-key, allows them to put their side of the case, but coaches them to amend their behaviour so that they do not cause distress. Here, investment in management training is worthwhile, as resolution may require a level of psychological tight-rope walking at times together with commensurate social skills on the part of the managers seeking resolution. Personnel or HR staff may find themselves providing coaching to those managers who are at the front line of informal systems, and this would be time well spent.

Some of these informal systems that involve gentle intervention might be seen as mediation. In some cases this will be true, and some organisations have tried to set up specific mediation services where a trained group of mediators can be brought into any situation by anyone who thinks that their services could be of use. Such services are being experimented with in workplace bullying, but the results are not clear yet. Unfortunately, when mediators are brought in, they often find that the situation has already escalated to such an extent that it is irretrievable, and their skills remain unused.

Being brought in 'too late' to resolve a situation is a major complaint of trade unionists also. The authors have considerable contact with trade unionists and join with them in reflecting how, once the conflict has escalated, the parties become entrenched beyond an informal resolution. This is particularly the case for the workplace representatives who will often know those involved, and who understand the negative role of conflict escalation and how it can change relationships from a tenable give-and-take situation to untenable demands and counter-demands.

By now we hope that readers will be realising the importance of early resolution to bullying at work and the commensurate importance of the role of informal systems. Quite rightly, formal systems are in place to deal with official complaints, but these formal complaints are often the outcome of a situation that has escalated too far to be resolved easily. Effective informal systems can 'pop' the conflict escalation balloon, negating the need for very expensive formal systems that can (in themselves) be damaging.

Formal processes

Most organisations that deal with bullying have a statement in their policy which indicates that bullying can be treated as a dismissal offence. Most organisations make bullying part of their description of 'gross misconduct' and address it in a 'normal' way.

This section will not outline all the possible approaches to formal complaints systems as that is not the purpose of the book. However, it will highlight how bullying can fit rather awkwardly into some existing systems which may need amendment in order to include bullying within the normal processes.

Formal complaints

In many formal systems, it is their line manager to whom a target should complain first. If reporting to the boss is the only route, and given that so many bosses are reported to be the bullies, this is clearly an inappropriate system. Thus one needs to have established second channels for the reporting of complaints. Here we should point out that the department which usually suffers most from poor systems of this kind is personnel or human resources. Often everyone in the organisation has HR or personnel as a 'second' line of complaint, but to whom do the staff within HR or personnel go if they are being bullied by their boss? One of the signs of a well-thought-through system is having two lines of complaint, one specifically for personnel and HR staff.

Investigations

Investigation processes are another area where bullying at work has presented challenges for systems and procedures. Frequently, the core of a complaint about bullying at work is a set of experiences which, when taken on their own, may seem rather innocuous, but when put together form a pattern of consistent undermining behaviour. Thus the investigator needs to have solid investigative training and skills.

Some formal systems only allow for evidence to be taken from the complainant and the alleged offenders. This is often insufficient for workplace bullying as it can result in two very different views on the same set of events. It is important to broaden the information-gathering process to people other than those directly involved in the complaint, for example witnesses or colleagues. Often people report being bullied with others, so in theory, getting collaborating information is possible. The investigator needs to remember the climate of fear that bullying can create. Fear of involvement and retribution can mean that people are not willing to make statements. In such cases, the investigator will need to provide enforceable assurances that those giving evidence will not experience later recriminations (Hoel and Cooper 2000a).

It is important that the investigator does not presume judgement before the evidence is collected. Such impartiality is crucial, and, while this may seem an obvious point, the emotions of those involved can be quite compelling and provide an extra stress on the investigator. The most recent areas of concern expressed by trade unionists and other staff representatives is that the alleged *bullies* need to get a full and fair hearing. As hard as bullying is to prove, it can be just as hard to disprove. Usually we will be dealing with judgement calls, and transparency and clarity are paramount.

Effective formal systems will have a time period within which the investigation must be completed. Unfortunately, many organisations with time limits do not adhere to them, as the situation is typically complex and requires more time than a regular investigation. Whatever the situation, it is important that those involved remember the reason for fast and timely systems: they help to minimise damage. If the investigation is so complex that it cannot be completed in the time available, it is important to let all involved know why.

Sometimes staff cannot be freed up to undertake investigations at short notice, and at other times it is hard to find staff who are at a sufficiently senior level to complete an investigation. For this reason several organisations employ external investigators to help them. We would support those who go down this route as often it is a highly efficient (and ultimately often less expensive) route. It is also an effective way to train new in-house investigators.

What happens to the target and the alleged perpetrator(s) during the investigation? For serious complaints, many organisations suspend all parties on full pay, asking them not to have any contact with those at work so that an investigation can progress without their influence. This is good practice, as investigations where the parties are still at work can be made impossible by lobbying or threats from either party to those who may be called to give evidence to the investigators. A key point, however, is not to underestimate the trauma of sitting at home and waiting for the investigation. Counselling is a sensible facility to offer those involved as it should reduce the feeling of isolation and may provide damage limitation during this difficult period. One error made by several British organisations is offering counselling to the targets but not to the accused. It is critical to treat everyone the same.

There will be a cost of providing counselling and other assistance to those who are sitting at home waiting for investigations to be completed. This would be offset by the cost of even one complaint against the organisation for inflicting harm *during* the official complaint process. This might seem harsh to some readers, but several organisations in Britain have been threatened with action as their process of investigating and resolving complaints has caused further harm to their staff. One can see such actions as further symptoms of a conflict which has escalated beyond the bullying incident. The threat of further court action from bullies is the reason why some of them are 'paid off' to leave the organisation rather than being dismissed.

Results of investigations

If bullying is found to have taken place, the organisation must take further action. In cases where the behaviour has been seen as gross misconduct the bully or bullies may be dismissed. Often, though, the situation is not this serious and formal warnings are issued. What happens to the bullies and the targets after the event? As we have seen before, the parties have been at a point of unresolvable conflict for some time and the idea that they should get back to working together can seem untenable in practice.

If either party is to be moved within the organisation, often the more junior people are relocated as they are easier to move. However, as more junior people are often the complainants, they may feel punished further by having to change their jobs. Equally, moving people might not be an option for those who work in smaller organisations as there is no position to which they can be moved.

This problem is a tough one for managers to resolve as the business may suffer if staff are moved. Usually managers value their more senior managers, and wish to see them continue in their present positions, but the fairness of this is likely to be questioned if those senior managers are the bullies. We would suggest that there are high costs involved in *not* moving a person if they are found to be a bully. If someone is allowed to continue in their job this can encourage others in the organisation to think that 'bullies can get away with it' (UNISON 1997). The organisation must be seen to act against bullying, otherwise staff will judge the organisation as supporting bullying.

Some personnel officers have reported that they use the internal grapevine to let it be known that warnings have been given and that bullies are being watched closely from now on. Once again we return to the problem of confidentiality. Personnel officers should not share the outcomes of a case, but they sometimes can see benefits to informing other staff that 'justice has been done'. This is a very tricky situation, and we can see the compelling arguments for letting the outcomes of the case be known. The authors have to remind readers that personnel proceedings are confidential and should be kept that way. In the short term, this might mean that those staff acting against bullying may do considerable work without being able to talk about it, but in the long run the climate *is* likely to change. They are indeed the un-sung heroes!

Summing up policy and procedure

Our discussion of policy and procedure demonstrates that formal and informal processes can help to resolve specific situations of bullying at work. We have made the argument that it is worth investing time and financial resources in developing effective informal processes so that the very expensive formal processes can be avoided and resolution can be found before tough decisions are forced upon everyone. We will now turn to the wider issues that are beyond the detail of the complaint and investigation processes and that can help establish and protect a non-bullying work environment.

Beyond policies: tackling bullying in other ways

As readers' concerns will vary, this section has been divided in order to reflect different levels of interventions that may be necessary or appropriate. We begin with those who have a situation of endemic workplace bullying and move towards those who are seeking to continue prevention and minimise the problem. It is unlikely that bullying can be eradicated altogether, and readers would be sensible to expect that bullying is going on in their organisations at a much higher rate than they may think. This statement is made partly because we believe this is the reality for most readers, but also because it is just sensible to take a prudent approach.

Tackling endemic bullying in your organisation

All interventions should start at the top, but if bullying is endemic it is likely that the senior managers have survived and participated in a bullying system during their own employment. This situation may present several problems. Senior managers or directors may be unwilling to label situations or cultures as bullying. They may be unwilling to devote resources to anti-bullying programmes, and they may play down the effects that bullying can have and the damage it can cause. In all instances, hard data can be used to counter these claims. However, it is unlikely that senior staff in such a situation will be enthusiastic about running a survey on bullying in order to get local evidence. We would encourage those who wish to gather data on the topic to be imaginative in their approach. For example, to include bullying as a small part of a general staff satisfaction survey might be the way to get these gatekeepers to agree to such an initiative.

Questions can be asked about bullying without directly using the 'bully' word. For example, you could take some of the behaviours shown in the UMIST study and include them amongst questions that you ask. The results can then be compared with those in this text to infer the level of bullying in your organisation. Staff might also be asked to deny the presence of positive behaviours. For example, questions that ask 'My manager is a pleasure to work with' or 'My manager can be trusted with personal information' can be quite revealing.

Beyond this, one can also present the financial case for dealing with bullying at work. Investigations can cost enormous sums in terms of management time – tribunals even more so. One can also calculate the cost of replacing staff who leave organisations because of bullying. Again, the data presented in this book can be used – 25 per cent of people leaving because of bullying is a robust number in the UK – for those outside the UK, other studies or academics in your own country may be able to identify the correct numbers.

We have already seen how policy and procedures can be ineffective, so one should not underestimate the ability (conscious or otherwise) of senior managers to undermine the process! One of the most undermining systems we have observed is what the authors call 'vacuum management'. This is where decisions fail to be made. Senior managers may hope that by not giving an answer, the situation will go away, and sometimes it does. Of course many staff will know about their tactics. Unfortunately this lack of action can lead to staff having no confidence in senior managers' willingness or ability to combat bullying at work, and unfortunately in some instances this is very well founded. How does one challenge those at the top? Eventually vacuum management undermines the competitiveness of the organisation because staff leave. However, this takes time to work through and bullying will probably be tackled after it is too late.

Effective managers will always be on the look-out for ways of reducing cost and certainly reducing bullying presents a terrific opportunity. However, managers in organisations where bullying is endemic are sometimes ineffective managers, so such arguments are unlikely to sway them.

Where bullying is endemic, the revolution needed has almost no chance of starting at the top. Instead, if one is going to do something about it, then changes in behaviour will need to start at the middle and at the bottom of the staff structure. As has been seen in the previous chapter, some staff will take action on an individual basis. Very strong individuals who have a strong sense of self-worth, the worth of others, and are willing to stand up in support of decent treatment will take such initiatives. If you are one of those individuals or thinking you would like to start being one of them, then it is important that you gather 'like' individuals and informally support each other. People in positions of leadership through their roles such as safety managers, HR professionals, personnel staff or trade union representatives may find that within their remits they can legitimate and initiate changes to affect bullying at work.

Support can be usually be gained for an anti-bullying policy (how does one present an argument against it?) and if the formation of the policy can include others, then it may prompt more staff to consider what behaviour is acceptable at work. We would have liked to suggest that a policy could provide a vehicle for a complete change in behaviour but that would be naïve. However, a policy can provide a helpful support structure for those who promote good behaviour at work. It can also circulate a positive message within the organisation which may present opportunities for the involvement of some staff. It can act as a vehicle for those who are able to challenge senior managers (such as trade union officials) and provide them with legitimacy for argument.

Co-ordinating work on policies can provide a meeting for like-minded people. For example, personnel, HR or occupational health professionals working with unions and staff associations on bullying policies can raise awareness and get more people thinking about their actions and the effect of negative behaviour. In such ways the culture can be seen to change. Getting all representative groups on board is to be highly encouraged, but it is important not to let any one group dominate, as the policy needs to be 'owned' by the whole organisation and not be seen as just a trade union or staff association initiative.

Fundamentally, if bullying behaviour is rewarded and no punitive measures are taken against it, one is fighting an uphill struggle. If the existing formal rewards and punishments do not act as a deterrent to bullying, we would suggest that you look at the informal systems. Reviewing Chapter 5 on culture may provide a framework for finding some appropriate angles. For example, we would encourage you to think through how 'power' works and how the acceptance (or rejection) of others might be used as an informal system of reward and punishment. We would encourage readers to act as role models for others and try to take any and every opportunity to demonstrate that it is possible to go about one's business without using bullying behaviours and, instead, treat others with dignity and respect.

Do use the policy when you have it in place. Make the policy public and make sure that the early complaints are dealt with in a very thorough manner. This will set the standard for behaviour and embed a system that can be trusted. It may be necessary to use outside consultants (for example to undertake investigations) to ensure this. Their brief may include trying to promote the policy to senior managers and coaching them in the application of the principles of the policy. The authors have also seen how outside consultants provide senior managers with a variety of legal scenarios which may frighten them into action in order to avoid negative publicity at tribunals and other public proceedings. While we would prefer senior managers to adopt the policies from a more positive point of view, warning of the dangers may be the only impetus initially available with staff at a senior level.

Organisations where bullying is endemic need to use systems that can be trusted. An unsophisticated system that works is better than a highly sophisticated system that falls at the first hurdle. Failure will be expensive as it will give ammunition to the sceptics of the programme. For example, if you cannot set up the necessary training for an effective contact system, don't set one up in the first place.

One reason we suspect that senior managers avoid taking on bullying at work is that they find dealing with it difficult and unpleasant, which it can be. While these reactions are understandable, avoidance of dealing with bullying can be another example of 'vacuum management' that enables bullying to survive and prosper. Some senior managers may not dare to confront key staff who may be accused of bullying. There may also be some senior managers who do not want to expose their own belief in doing business at any human cost. Other senior managers may find it too hard to take actions against their friends and colleagues of a number of years. All of these are understandable reasons. Dealing with bullying at work does test leadership abilities, and some managers simply find the decisions too hard.

We have observed two major patterns of management failure in dealing with workplace bullying, and both involve a lack of action. The first type is where staff at a senior level in management fail to support those further down the hierarchy who are trying to deal with the issue. Such failures range from not supporting policies generally through to not making appropriate decisions regarding staff who are found to be bullies, all of which effectively undermine anti-bullying programmes. The damage caused by this type of management failure can spread across the organisation as people judge their leaders' actions, particularly their attitude to the anti-bullying measures.

The second pattern of failure is at a lower level where local managers disregard action taken to combat bullying, perhaps by denying the existence or importance of bullying cases. A lack of action at this level is another form of undermining. Usually the effect of such action is only localised in the short term, by lowering the morale of staff and confirming their negative views of management. However, if situations are allowed to continue, they may escalate and in the long term lead to higher-profile situations if targets decide to pursue further action. If the two types of management failure exist together (as is often the case when bullying is endemic in an organisation), then any change will rely on individual action by individual managers as effectively there is no support for anti-bullying activity.

Monitoring

Readers are encouraged to think through carefully how they monitor bullying at work. For those working in situations where the bullying is endemic it is important for them to judge how their efforts are paying off. We would argue that monitoring is essential to any process connected to negative behaviour at work.

What should you monitor? All forms of absence should be monitored, including days taken as sick-leave, voluntary days, exit rates and any other form of absence such as leave without pay. Our data would point to long-term sickness absence and high exit rates as two key indicators for bullying at work. The higher these are, the more suspicious one should be that bullying is going on. Interviewing staff who have left the organisation has been a highly effective way of picking up problems and identifying pockets of bullying that revolve around certain individuals.

Other obvious areas for monitoring include the use of formal processes to deal with bullying, and their outcomes. One might see increases in the use of procedures at the start of a programme, but over time these should diminish. Gathering data on the use and outcomes of schemes should be fairly straightforward and initiated at the start of the programme.

The gathering of data on the use of informal systems and procedures is less straightforward, and the less formal the process, the more difficult it is to measure. Several structured, informal systems for assisting targets of bullying such as contact adviser schemes and employee assistance programmes, have been mentioned. The use of such systems should be monitored in terms of the number and nature of enquiries and their outcome. Further data-gathering will depend on the level of confidentiality which has been offered to users of such schemes, but the pattern of enquiries, in terms of which departments and sites they come from, might be able to distinguish the corporate situation as having either endemic bullying or pockets of bullying. The problem with such data is that low numbers may imply a lack of trust in the scheme itself rather than low levels of difficulty being experienced by staff. It is worth monitoring such schemes independently, for example by asking staff via interview or questionnaire about their awareness of, and confidence in, the scheme.

We would encourage readers to gather as much information as possible on informal complaints (not only from contact advisers) and assemble the information so that a full picture can be generated. Staff representative organisations such as trade unions might come in here. Staff that are both central and peripheral to complaints about bullying should be canvassed and these might include occupational health service staff, safety representatives and personnel. More information can be gathered through talking with such staff rather than by written methods, as considerable information may be passed on 'unofficially' which they might be unwilling to commit to paper.

It is quite easy to have a situation where different complainants are going to different staff about the same bully or bullies and which is not picked up on at the centre. This is a real danger in larger organisations or those organisations in which staff can turn to a variety of people for help. One must pull together all data on informal systems in order to be able to monitor effectively. Collating the data may reveal patterns of complaint after which further actions may be taken. It is at such a point when the inclusion of someone senior in the anti-bullying programme would add sufficient authority to ensure changes were made.

So far, all the evidence considered can be gathered from people who are involved in the anti-bullying programme. We would strongly urge you to take a much more proactive route. Annual surveys of staff opinion are increasingly seen as good practice in the UK, and we welcome such initiatives. Again, this is harder for the smaller organisation, but they might use external agencies to undertake straightforward monitoring. The inclusion of questions that relate to ordinary behaviour from managers and colleagues is important, and such surveys can provide annual benchmarks for changes in attitude and opinion. The situation may appear initially to worsen. This might be because awareness is raised and people reconsider and re-classify the behaviour they experience. In the long term, falling rates of negative behaviour and higher retention levels should be evident.

Monitoring should be regular, for example every six months, otherwise impetus can be lost. The monitoring process should have a clear structure, and the use of soft information such as feelings and opinions as well as the harder information of system use and categorised outcomes should be employed. The information should be reviewed at least once a year by a team who then can take decisions regarding changes in the schemes.

Finally, the monitoring system itself should be audited regularly. This is perhaps best done by external advisers since this will create a stronger validity to the outcome of their results (as they have less conflict of interest than internal staff). It will also provide other ideas for improvement beyond the internal team. To some readers this may appear to be a case of monitoring 'overkill', but there are two interconnected reasons for auditing the monitoring process. First, if the monitoring process itself is defective, then the evidence on which any changes may be made to the programme may also be flawed. This can lead to inappropriate action being taken. Second, if the organisational processes regarding bullying at work are scrutinised for any reason (for example at a court case), then such diligence will demonstrate that the organisation takes this issue seriously – but only as long as you have reacted to the findings of the audit.

Tackling 'pockets' of bullying

Bullying behaviours at certain sites or in certain areas of the organisation may be of concern. When investigating pockets of bullying we suggest that you go back to Chapter 5 and use the culture web to diagnose the issues at that site or in that function.

A key issue here is to discover what is sustaining bullying at that site or within that group of people. Straightforward questions should be asked. What is different in these sections? What is rewarding bullying behaviour? How can you undo such rewards? Consideration should be given to whether the situation reflects a 'rotten apple' (i.e. an individual manager) or instead reflects a 'rotten barrel' which indicates something deeper in the system. Each requires a different intervention – changing the manager in the first instance, or changing reward systems or attitudes and expectations in the latter 'rotten barrel' situation.

Prevention of bullying

Some people will be reading this book so that they can keep up the excellent standards of behaviours within their organisations and prevent bullying taking a foothold. For prevention, monitoring is critical, especially gathering data on why people leave and exit rates.

Rewarding positive behaviour towards others is also essential. Organisations where bullying is at a very low level are likely to be highly proactive. They will be walking the corridors on the lookout for unhappy staff, *and* engaging with them. Readers are advised to retain that level of honesty and keep transparent levels of appraisal and awareness. Managers in such organisations do not hide away, but make it their business to continue walking the corridors and the shop floors so that they gather base data from ordinary staff and not rely on data filtered by others. They will keep assuming that they have a problem. They will examine any changes in reward systems and be particularly carefully to ensure that they do not encourage bullying.

And finally . . .

As we have seen, once an organisation has a policy with the accompanying procedural systems for informal and formal complaints, it will need to be monitored. The results of the monitoring will also need to be acted upon. The final step is to audit the process to ensure it is acting well.

The authors expect that the best organisations have a range of informal systems. We also expect that the best organisations have managers and colleagues who are always on the look-out for negative behaviour and have a clear idea on what they should do if they come across it. We expect that the best organisations have had very little cost associated with bullying at work because they are dealing with it early rather than having to pursue expensive internal investigations and highly costly tribunals. We also expect that the best organisations have low exit rates and highly committed staff. Finally, we also think that all these issues are connected.

11 Future developments

Many ideas have already been presented which could help us further our understanding of bullying at work. In this final section of the book we would like to reflect briefly on some of these ideas, how we can develop them, and the potential challenges they present.

In the UK, progress in this field since 1995 has been remarkably quick. The authors have paid credit to the press for raising initial awareness of this issue. Subsequent research has been made more efficient by practitioners and researchers working together in partnership. The base data in Britain has been gathered via collaboration between academics, trade unions and employer groups. For example, Hoel and Cooper's (2000) national study was a co-operation between a number of employers, unions and academics. The UNISON studies showed similar good practice in collaborative working. For the academic, the involvement of those who are 'at the coal face' enables the research to be tuned appropriately to current and relevant issues. Hopefully, academics bring ideas and conceptual contributions that enhance a study beyond one that might be designed 'in-house'. As our experience grows, so the strengths and weaknesses of different designs are tested and built upon, and so our knowledge also grows.

Practitioners are demanding of academics. The pressure is on for us to contribute to this issue. At first academics were asked to provide evidence of the extent of bullying, and to help with how it should be defined. Then academics moved on to constructing the argument relating to cost. Our focus has now moved on to the assessment of interventions and the provision of models of good practice.

The focus on interventions has changed the challenge for future work and has implications for both the nature of our enquiry and how we undertake it. Surveys have been very helpful in providing the base data for establishing the extent and nature of bullying, but at this stage we are really only able to describe it. This book has made connections relating to a deeper understanding of bullying at work so that we can progress to *explaining* it, rather than simply describing it. If we are going to tackle the intervention issues, we will need to have a better understanding of what is going on so that we can, in turn, understand why certain interventions work well.

Thus our future work is likely to move away from large surveys to qualitative work involving interviews and case studies. Such methodologies can gather information of a richer kind from which we can build theories and then test them. We should not be too optimistic about clear results. As we have already shown, bullying at work involves multiple layers and multiple analyses, so trends and general indications are probably the most realistic outcomes to expect. As with the ideas in this book, practitioners will then have to assess their own circumstances to see which are most applicable to their organisation.

Diversity

Central to the question of intervention are issues connected to diversity. How far can (or should) the organisation prescribe behaviour? Even within a single-nation group, the effort to describe 'acceptable behaviour' is extremely hard. When the scope extends to a multicultural and multinational context, the complexity multiplies. Fortunately diversity is already on the agenda of many practitioners, and we would encourage the inclusion of bullying and negative interpersonal behaviours in their deliberations.

It may not be coincidental that diversity and bullying at work should be appearing on our HR agendas around the same time. The challenge of management may have changed as we progress toward a working environment which is more demanding on all sides. Employers need more and ever-higher quality workers to remain competitive and employees expect decent treatment in return. The authors suggest that our training of managers needs to better reflect the pressures that they experience, and we expect our future work to encompass such initiatives.

Hands up, bullies

We have insufficient information about people who bully. Existing data points to a plethora of possibilities which includes (at a descriptive level) single operators, ringleaders or followers. Typologies such as that employed by Marias and Herman (1997) and shown in Chapter 4 encompass many possibilities. Paucity of data from those who have been identified as 'bullies' themselves negates any conclusions regarding bullies.

Once again, let us step back and think about the identity of the 'bullies'. If we consider that half the population will have contact with bullying at some stage in their working lives, there must be a great number of bullies about. We also know that bullying seems to affect all layers in the hierarchy, men and women, regardless of age.

If we seek to reduce the damage of bullying at work, then we must seek to reduce the experience of negative behaviours at work. Who is behaving negatively? For example, how many of us can look at the list of behaviours from Hoel and Cooper's study and say we have not done some of these in the last year? Energy employed searching for the 'demon bully' might be better spent looking closer to home.

By not demonising the bullies, we hope to encourage ordinary people to examine their own behaviour at work. We would prefer to adopt an approach from racism and sexism training which begins by exploring prejudices with regard to race and sex in all of us. Bullying at work appears to be just as prevalent as sexism and racism, and perhaps requires an approach that allows us all to examine the less pleasant side of our behaviour at work.

It is unlikely that 'How Not To Be A Bully' will ever be included in our corporate training programmes! However, we would encourage readers to make every effort to include negative behaviours at work within ordinary training sessions. Promoting good practice and respect for each other is essential, but negative behaviour *is* seen and judged by others and needs a profile in training too.

Organisations that bully

Whilst all individuals should take responsibility for any elements of their behaviour that are experienced negatively by colleagues, we must also acknowledge the role of management and the organisation in bullying. We anticipate that the tension between individual responsibility and corporate responsibility will increase. Some people find it easy to blame their organisations as being 'a bully', and this may prevent them from acknowledging their own behaviour. That said, the organisational context can be a real problem.

If the organisational culture accepts bullying, then individuals who try to combat it face an uphill struggle. We hope that we have presented evidence that bullying creates a huge drain on resources by, for example, staff leaving and taking time off in sick-leave. If readers were to estimate that 10 per cent of their staff are bullied, that 25 per cent of those leave, and then multiply this figure by an average replacement cost, one example of direct cost will be found. After that, another cost can be estimated for witnesses (around 20 per cent leave). The costs in these areas alone are large.

When dealing with bullying, professionals need to be careful that they do not take on the role of bully. Targets who go to court frequently state that the process of seeking legal redress is as traumatic as the bullying incidents themselves. Those responsible for the design of internal investigations should be mindful of this and work to protect all staff (including the alleged bully or bullies) against unnecessary harm and further damage. Appearing not to listen to targets and providing inadequate support for those accused of bullying are common stories. Practitioners are under the spot-light themselves and *being seen* to be fair is fundamental to the many staff who will be aware of and involved in the process.

At a broader level, those staff in management positions who agree to clearly unachievable targets might think twice about their employer. Being set up to fail happens at all levels, and, like bullying at the interpersonal level, it is a potentially destructive conflict. However, one can see little room for movement by the executive who works in a climate where the discussion of 'failure' or 'non-achievement' may risk their job security. Such denial only protects those higher in the hierarchy from the pain of a reality that may come crashing down around them in a short while. Such negative dynamics exist, and are extremely hard to change.

We should also take a moment to consider the situation of people working in small firms. Whilst this book has tried to present good practice for organisations of any size, when an owner manager is a bully the options are limited if an initial approach highlighting their behaviour is rejected. Smaller organisations often have weak union representation, so little external help or pressure is available. Other than trying to work through people whom the owner manager respects, the authors can only suggest the employee reconsiders their position. In these instances, the 'bully' and the 'organisation' really might be the same.

We also have concern that organisations could be seen to 'bully passively' by failing to act or respond. In Chapter 2 we saw that people who were bullied reported 'nothing' as the most common outcome of contact with professionals. We should therefore remind ourselves that doing nothing is not a neutral act in bullying at work (Rayner 1998), but rather, as shown in Chapter 8, is often interpreted as collusion with bullying. It is very likely that professionals will have taken action, but confidentiality rules prohibit them from disclosing to targets. Confidentiality was discussed in Chapters 9 and 10, and it is emerging currently as a critical issue.

It is important for organisations to find a way to tell their staff that they are, indeed, taking action about bullying at work. This could be fairly straightforward in a large organisation where annual figures for complaints and outcomes could be published. It would be harder in smaller organisations as confidentiality may be compromised. This said, the authors would strongly encourage more thought and creativity to be put into giving feedback to staff on outcomes, otherwise there is a risk of being seen as colluding with bullying at work.

A final comment on the bully is to question the role and use of performance indicators such as league tables and other measurement indicators. These systems create situations where there are a few 'winners' and many 'losers'. English education league tables have been used as a catalyst for a case study in Chapter 6. In the UK we are now seeing 'Beacon Council' status as the new league table device for local authorities. Those responsible for developing and putting in place such ranking systems need to consider how bullying at work might be encouraged through the use of such systems and also the actions necessary for minimising their potential negative effects.

The ripple effect

Survey evidence points to the potential for damage to those who have witnessed bullying at work. Many people leave organisations after witnessing bullying. This 'ripple effect' needs further and urgent investigation as it represents another major cost to organisations. In addition, we have also seen little research with those who are involved in a professional capacity, for example as supporters of targets or investigators. This is unlikely to be a stress-free area. Such employees will be mindful to keep themselves as personally unconnected as possible, but, in practice, it must be extremely hard at times. It would be useful to find out the nature and extent of this type of 'ripple' and to identify examples of good practice in minimising and dealing with the effects.

As researchers we would also like to see our scope of investigation extend beyond those involved in the workplace where the bullying is taking place. This broadening might include friends, family, neighbours and anyone who has contact with targets or witnesses. Anecdotal evidence reveals how targets react very differently to being bullied, with some people talking about it a great deal and others keeping very quiet (e.g. Adams 1992). However, those whose advice or listening ear is sought may find their situation difficult, and are less likely to have the training or the coping capacities of those who deal with bullying as part of their job.

We also suspect the existence of a further 'ripple', namely the development of a negative reputation for the employer locally. If this were the case, finding staff may become a problem and dealing with other negative public relations issues may prove problematic as there will be little sympathy in the community. These are only suspicions, however, and proper evidence is needed to decide whether such phenomena should be added to the 'cost of bullying' bill.

Some studies have widened the concept of 'bullies' to include those who do not actually work for the organisation. Whilst this might include customers of a corporate nature, it is most likely to be clients of service organisations. Already there is evidence from hotel and catering staff and those who work in the NHS that clients are the source of some bullying behaviour (e.g. Quine 1999; Hoel and Cooper 2000). Clearly this may affect many sectors and needs further work. This development would challenge the current concept of bullying as being between people who know each other over a period of time.

Certainly some organisations are taking up the issue of protecting their staff against negative behaviour by clients. At the time of writing, the entrance to the customer lounge of the British Airways terminal at Heathrow has a large notice which clearly indicates that negative behaviour towards staff will be dealt with. This is a positive development, as some firms have taken the attitude that the customer is 'always right' to the extent of ignoring mistreatment of staff.

The nature and prevalence of bullying at work

By now, we would expect there to be sufficient data available to convince the sceptic that bullying at work forms a significant workplace issue in the UK. Why only half those who experience bullying behaviours actually label themselves as bullied presents an interesting problem. Investigating why people label themselves (or not) could provide some valuable insights into the dynamics at work in the bullying process. In some cases, if staff labelled their experience, we could help them faster and easier. A shared idea that something is wrong would exist, and an awareness that action needs to be taken. Knowing why people do not label themselves may help us to understand some of the myths around bullying at work, for example, some people may find it to be a protective measure in order to resist the role of the 'victim' (Einarsen and Hellesøy 1998). This area is high on research agendas currently.

We have started to understand the complex dynamic of escalating conflict which results in people being bullied at work. We need further insight into how such events escalate unchecked. Anecdotal evidence both from personnel and trade union representatives points to 'bullying' situations that are frequently past the point of mediation services. Not all conflicts produce bullying: what is different about these circumstances? If we are to look at the behaviours, we have to see them in the context of their dynamic. To be able to do this, we need to have a better understanding of the nature of the dynamics.

How are we trying to understand bullying at work?

The 'problem' might be left with the targets of bullying as their work status and health erode, but it would be erroneous to see the 'solution' as residing with those individuals. Our scope needs to extend beyond the targets to the bully or bullies and the workplace itself. The problem of bullying at work cannot be understood at any one single level. Rather it is an amalgamation of multiple levels – those of the individual, the work group and the organisation, for example.

In understanding bullying at work we often assume that there is one 'truth' (Hoel *et al.* 1999). For example, I may have the experience that someone is 'out to get me' and that I am being bullied. An independent person might question the alleged perpetrator and discover that they are not 'out to get me', but that they can see how I might have reached that conclusion. In this case, the *experience* of someone being 'out to get me' is true, but the *reality* is that they are not. In such cases, 'facts' are ambiguous and their use demands care and sensitivity

Anyone with experience of investigating cases of bullying at work will know this dilemma well. Investigators of bullying can frequently see how 'reality', constructed from one's individual position (and perhaps reinforced by colleagues, friends and family) can be at odds with someone else's point of view. Stories, questionnaire responses and witness statements all represent opinion. However, as shown in Chapter 10, such stories can take on a 'life of their own' and deviate from the 'truth'.

When thinking about bullying at work, there is a tension between the 'truth' on the one hand and the subjective nature of bullying, which in turn is also a 'social construct', on the other hand. In the experience of the authors, practitioners need little encouragement to focus on the tangible and real. However, practitioners need to be aware of how their staff judge behaviours, interactions and messages of bullying at work because these interpretations can affect what they and others perceive as the 'truth'.

Moving forward

In order to move forward, researchers and practitioners need to continue their collaborative approach. Wherever you are reading this, we would encourage you to continue to talk about the issue. Raising the issue of bullying at work, like sexism and racism, does not encourage nor inhibit it, but starts us on the way to dealing with it.

Appendix 1
Sampling for surveys

The following is an example of instructions given to organisations that wish to sample for studies such as those which investigate bullying at work.

Company A is requested (if participation is agreed) to identify a total sample of 100.

This sample should be made up of employees from the following groups:

All categories of staff

It is important that the individuals selected for participation in the study are identified randomly. This means that no consideration should be given to the history or qualities of individual participants. Such a sample can be computer-generated from a database of employees. The best way forward would be to create a computer file, sorting employees alphabetically by surname and then programme the computer to identify every fifth person (based on a total number of employees of 500). This should then give you approximately the correct sample number and ensure that the sample is random as well as representative. A simpler way of doing the sampling would be to select every fifth person on a list of employees.

If possible, the sample should reflect the shape or hierarchical structure of the organisation, i.e. you should try to ensure that the selection method reflects the size of the various groupings within the organisation, e.g. fewer supervisors than operators or ordinary workers, and so on. It is also important to get a reasonably good spread as regards sex, age and seniority/years of service. I would recommend that you initially select the sample totally at random and on the basis of the selected sample, check for the other variables. If the random sample is very much skewed, for example representing only 25 per cent females whereas in reality every second person is a woman, you may have to replace some male employees with some more women selected at random. However, whilst it is important that you pay some attention to these issues, as it will affect results, we do not expect you to spend a lot of time and effort trying to meet the demand for representativity down to the smallest detail.

Source: Hoel and Cooper 2000.

Appendix 2
Negative behaviours at work

Experience of negative behaviours (in percentages) – total sample

Exposure to individual negative behaviours in ranked order – questionnaire item number	Never (%)	Now and then (%)	Monthly (%)	Weekly (%)	Daily (%)
Someone withholding information which affects your performance 1	32.7	46.9	7.1	9.1	4.2
Unwanted sexual attention 2	89.7	8.7	0.4	0.7	0.5
Being humiliated or ridiculed in connection with your work 3	68.6	25.4	2.4	2.4	1.2
Being ordered to do work below your level of competence 4	54.2	31.3	3.8	4.7	6.0
Having key areas of responsibility removed or replaced with more trivial or unpleasant tasks 5	61.7	28.6	3.4	3.1	3.0
Spreading of gossip and rumours about you 6	66.1	26.5	3.3	2.2	1.9
Being ignored, excluded or being 'sent to Coventry' 7	80.9	14.2	1.8	1.7	1.4
Having insulting or offensive remarks made about your person (i.e. habits and background), your attitudes or your private life 8	75.4	17.9	2.2	2.1	2.4
Being shouted at or being the target of spontaneous anger (or rage) 9	70.3	22.2	2.9	2.8	1.9
Intimidating behaviour such as finger-pointing, invasion of personal space, showing, blocking/barring the way 10	82.5	12.3	1.6	2.2	1.5
Hints or signals from others that you should quit your job 11	88.9	8.7	0.9	1.0	0.5

Exposure to individual negative behaviours in ranked order – questionnaire item number	*Never (%)*	*Now and then (%)*	*Monthly (%)*	*Weekly (%)*	*Daily (%)*
Threats of violence or physical abuse 12	89.6	6.7	1.4	1.5	0.9
Repeated reminders of your errors and mistakes 13	72.0	23.5	2.3	1.3	1.0
Being ignored or facing hostility when you approach 14	74.1	19.7	2.7	2.2	1.4
Persistent criticism of work and effort 15	78.7	16.2	2.4	1.9	0.7
Having your opinions and views ignored 16	42.8	42.9	6.4	4.9	2.9
Insulting messages, telephone calls or e-mails 17	92.4	5.6	0.7	0.7	0.6
Practical jokes carried out by people you don't get on with 18	92.0	6.7	0.6	0.4	0.2
Systematically being required to carry out tasks which clearly fall outside your job description, e.g. private errands 19	78.4	16.6	1.8	1.6	1.5
Being given tasks with unreasonable or impossible targets or deadlines 20	48.1	35.0	7.2	5.8	3.9
Having allegations made against you 21	82.5	14.7	1.6	0.7	0.5
Excessive monitoring of your work 22	72.7	18.0	4.2	2.4	2.8
Offensive remarks or behaviour with reference to your race or ethnicity 23	95.4	3.1	0.5	0.3	0.6
Pressure not to claim something which by right you are entitled to (e.g. sick-leave, holiday entitlement, travel expenses) 24	71.0	22.3	3.1	1.8	1.7
Being the subject of excessive teasing and sarcasm 25	83.9	12.2	1.4	1.4	1.1
Threats of making your life difficult, e.g. over-time, night work, unpopular task 26	87.4	9.2	1.7	1.0	0.6
Attempts to find fault with your work 27	74.5	20.1	2.8	1.6	1.1
Being exposed to an unmanageable workload 28	46.2	32.8	6.5	6.7	7.9
Being moved or transferred against your will 29	81.8	15.2	1.3	0.8	1.0

Source: Hoel and Cooper 2000.

Appendix 3
Useful contacts

A variety of organisations exist in the UK which may help signpost further help.

The Andrea Adams Trust
Tel: 01273 704901

The Suzy Lamplugh Trust
Tel: 020 8876 0305

Websites

Because domain names change, we have made some suggestions to use with search engines.

Good websites for information on bullying at work include Tim Field's site which can be searched for under 'successunlimited'. This site has excellent links to other pages around the world.

'Bullybusters' is a site run by Gary and Ruth Namie who are psychologists and campaigners in the United States.

Susan Steinman runs an informative site based in South Africa which can be accessed by searching for 'worktrauma'.

The Beyond Bullying Association, based in Australia, has a website which is worth visiting.

A site at www.workplacebullying.co.uk does not provide personal contact but has many signposting pages and links to other groups.

References

Adams, A. (1992) *Bullying at Work – How to Confront and Overcome It,* London: Virago.

Archer, D. (1999) 'Exploring "bullying" culture in the para-military organisation', *International Journal of Manpower* 20(1/2): 94–105.

Ashforth, B. (1994) 'Petty tyranny in organizations', *Human Relations* 47: 755–78.

Babiak, P. (1995) 'When psychopaths go to work: a case study of an industrial psychopath', *Applied Psychology – An International Review* 44(2): 171–88.

Baron, R.A. (1990) 'Attributions and organizational conflict', in S. Graham and V. Volkes (eds) *Attribution Theory: Applications to Achievement, Mental Health and Interpersonal Conflict,* Hillsdale, NJ: Erlbaum.

Bassman, E. (1992) *Abuse in the Workplace,* New York: Quorum.

Baumeister, R.F., Smart, L. and Boden, J.M. (1996) 'Relation of threatened egotism to violence and aggression: the dark side of self esteem', *Psychological Review* 103(1): 5–33.

Baumeister, R.F., Stillwell, A. and Wotman, S.R. (1990) 'Victim and perpetrator accounts of interpersonal conflict: autobiographical narratives about anger', *Journal of Personality and Social Psychology* 59(5): 994–1005.

Besag, V. (1989) *Bullies and Victims in Schools,* Milton Keynes: Open University Press.

Bjorkqvist, K., Osterman, K, and Hjelt-Back, M. (1994) 'Aggression among university employees', *Aggressive Behaviour* 20: 173–84.

Bosch, G. (1999) 'Working time: tendencies and emerging issues', *International Labour Review* 138: 131–50.

Bowles, M.L. (1991) 'The organization shadow', *Organization Studies* 12(3): 387–404.

Brown, A. (1998) *Organisational Culture,* Harlow: FT/Prentice Hall.

CBI (2000) *Focus on Absence: Absence and Labour Turnover Survey 2000,* London: CBI.

Christie, R. and Geiss, F.L. (1970) *Studies in Machiavellianism,* New York: Academic Press.

Cooper, C.L. and Payne, R. (1988) *Causes, Coping and Consequences of Stress at Work,* Chichester: John Wiley & Sons.

Cooper, C.L., Sloane, S.J. and Williams, S. (1988) *Occupational Stress Indicator Management Guide,* Windsor: NFER-Nelson.

Cowie, H., Jennifer, D., Neto, C., Angulo, J.C., Pereira, B., Del Barrio, C. and Ananiadou, K. (2000) 'Comparing the nature of workplace bullying in two European countries: Portugal and UK', paper presented at Transcending Boundaries Conference, Griffith University, Brisbane, Australia, 6–8 September 2000.

Cox, T., Griffith, A. and Rial-Gonzalez, E. (2000) *Research on Work-related Stress,* Luxembourg: European Agency for Safety and Health at Work.

Coyne, I., Seigne, E. and Randall, P. (2000) 'Predicting workplace victim status from personality', *European Journal of Work and Organizational Psychology* 9(3): 335–49.

Crawford, N. (1997) Unpublished keynote paper presented at the Bullying Conference, Staffordshire University, UK.

Crawford, N. (1999) 'Conundrums and confusion in organisations: the etymology of the work "bully"', *International Journal of Manpower,* 20(1/2): 86–93.

Daily Mail (2000) 'Psychopaths among us' by Christopher Matthew, p. 63.

Davenport, N.Z., Distler Schwartz, R. and Pursell Elliott, G. (1999) *Mobbing: Emotional Abuse in the Workplace,* Ames, Iowa: Civil Society Publishing.

De Board, R. (1978) *Psychoanalysis of Organization,* London: Routledge.

De Dreu, C. and Van de Vliert, E. (eds) (1997) *Using Conflict in Organizations,* London: Sage.

Denenberg, R.V. and Braverman, M. (1999) *The Violence-Prone Workplace,* Ithaca, NY: Cornell University Press.

Earnshaw, J. and Cooper, C. (1996) *Stress and Employer Liability,* Harmondsworth: Penguin.

Einarsen, S. (2000) 'Bullying and harassment at work: unveiling an organizational taboo', paper presented at Transcending Boundaries Conference, Griffith University, Brisbane, Australia, 6–8 September 2000.

Einarsen, S. (1999) 'The nature and causes of bullying at work', *International Journal of Manpower* 20(1/2): 16–27.

Einarsen, S. (1996) 'Bullying and harassment at work: epidemiological and psychosocial aspects', PhD Thesis, Dept of Psychosocial Science, University of Bergen.

Einarsen, S. and Hellesøy, O.H. (1998) 'Når samhandling går på helsen løs: Helsemessige konsekvenser av mobbing i arbeidslivet', *Medicinsk årbok 1998,* København: Munksgaard.

Einarsen, S.E. and Matthiesen, S.B. (1999) 'Symptoms of post-traumatic stress among victims of bullying at work', *Abstracts for the Ninth European Congress on Work and Organizational Psychology,* Helsinki: Finnish Institute of Occupational Health, p. 178.

Einarsen, S. and Raknes, B.I. (1997) 'Harassment at work and the victimization of men', *Violence and Victims* 12: 247–63.

Einarsen, S. and Skogstad, A. (1996) 'Bullying at work: epidemiological findings in public and private organizations', *European Journal of Work and Organizational Psychology* 5(2): 185–202.

Einarsen, S., Matthiesen, S.B. and Mikkelsen, E.G. (1999) *Tiden leger alle sår: Senvirkninger av mobbing i arbeidslivet*, Institutt for Samfunnspsykologi, Bergen: University of Bergen.

Einarsen, S., Raknes, B.I. and Matthiesen, S.B. (1994) 'Bullying and harassment at work and its relationship with work environment quality: an exploratory study', *European Work & Organizational Psychologist* 4: 381–401.

Einarsen, S., Raknes, B.I., Matthiesen, S.B. and Hellesøy, O.H. (1994a) *Mobbing og Harde Personkonflikter. Helsefarlig samspill på arbeldsplassen*, London: Sigma Foriag.

European Foundation for the Improvement of Living and Working Conditions (2000) 'Third European survey on working conditions', Dublin, Ireland (unpublished findings).

European Foundation for the Improvement of Living and Working Conditions (1996) 'Second European survey on working conditions', Dublin, Ireland.

Eurostat (1995) 'Can long working hours kill?', *Labour Research* 84: 21–2.

Farrell, G. (1999) 'Aggression in clinical setting: nurses' views', *Journal of Advanced Nursing* 29(3): 532–41.

Field, T. (1996) *Bullying in Sight*, Wantage, Oxon: Success Unlimited.

Fine, M. (1985) 'The social construction of "What's fair" at work', *Journal of Applied Social Psychology* 15(2): 166–77.

Fitzgerald, L.F. and Shullman, S. (1993) 'Sexual harassment: a research analysis and agenda for the 1990s', *Journal of Vocational Behaviour* 42: 5–27.

Magley, V.J., Hulin, C.L., Fitzgerald, L.F. and DeNardo, M. (1999) 'Outcomes of self-labelling: sexual harassment', *Journal of Applied Psychology* 84(3): 390–402.

Geen, R.G. (1990) *Human Aggression,* Milton Keynes: Open University Press.

Ghoshal, S., Bartlett, C.A. and Moran, P. (1999) 'A new manifesto for management', *Sloane Management Review*, Spring: 9–20.

Goldberg, D.P. (1978) *Manual of the General Health Questionnaire*, Windsor: NFER-Nelson.

Goleman, D. (1996) *Emotional Intelligence*, London: Bloomsbury.

Guardian (2000) 'Prison chief admits reign of terror by jail staff', 14 February.

Handy, C.B. (1995) *The Empty Raincoat,* London: Arrow.

Hawton, K. (1987) 'Assessment of suicide risk', *American Journal of Psychiatry* 143(12): 145–53.

Hoel, H. (1997) 'Bullying at work: a Scandinavian perspective', *Institution of Occupational Safety and Health Journal* 1: 51–9.

Hoel, H. and Cooper, C.L. (2001) 'Origins of bullying: theoretical frameworks for explaining bullying', in N. Therani (ed.) *Building a Culture of Respect: Managing Bullying at Work*, London: Taylor & Francis.

Hoel, H. and Cooper, C.L. (2000) 'Destructive conflict and bullying at work', November 2000, unpublished report, University of Manchester Institute of Science and Technology, UK.

Hoel, H. and Cooper, C.L. (2000a) 'Working with victims of workplace bullying', in H. Kemshall and J. Pritchard (eds) *Good Practice in Working with Victims of Violence*, London: Jessica Kingsley Publishers, pp. 101–18.

Hoel, H., Cooper, C.L. and Faragher, B. (2000) 'Organisational implications of the experience of persistent aggressive behaviour and "bullying" in the workplace', paper presented at the Academy of Management Conference, Toronto, 5–8 August 2000.

Hoel, H., Rayner, C. and Cooper, C.L. (1999) 'Workplace bullying', *International Review of Industrial and Organizational Psychology* 14: 189–230.

Hoel, H., Sparks, K. and Cooper, C.L. (2001) *The Cost of Violence/Stress at Work and the Benefits of a Violence/Stress-Free Working Environment*, Geneva: International Labour Organization.

Hofstede, G. (1980) *Culture's Consequences: International Differences in Work-Related Values*, Newbury Park, Calif.: Sage Publications.

Hogan, R. and Hogan, J. (1995) *The Hogan Personality Inventory Manual* (3rd edn), Tulsa: Hogan Assessment Systems.

Hogan, R., Curphy, G.J. and Hogan, J. (1994) 'What we know about leadership', *American Psychologist* 49(6): 493–504.

Hotel and Caterer (1995) Letter from Neil Savage, 12 October, pp. 40–42.

IDS (1999) *Harassment Policies*, IDS Studies No. 662, London: Incomes Data Services.

IPD (Institute of Personnel Development) (1996) 'One in eight workers are victims of bullying reveals new IPD survey', Press Release, 28 November.

Ishmael, A. (1999) *Harassment, Bullying and Violence at Work*, London: The Industrial Society.

Janoff-Bulman, R. (1992) *Shattered Assumptions: Towards a New Psychology of Trauma*, New York: The Free Press.

Jehn, K. (1994) 'Enhancing effectiveness: an investigation of advantages and disadvantages of value-based intragroup conflict optimising performance by conflict stimulation', *International Journal of Conflict Management* 5(3): 223–38.

Jensen, I.W. and Gutek, B.A. (1982) 'Attributions and assignment of responsibility in sexual harassment', *Journal of Social Issues* 38(4): 121–36.

Johnson, G. and Scholes, K. (1997) *Exploring Corporate Strategy: Text and Cases* (4th edn), London: Prentice Hall.

Jones, E.E. and Davis, K.E. (1965) 'From acts to dispositions: the attribution process in person perception', in I.L. Berkowitz (ed.) *Advances in Experimental Social Psychology* 2, New York: Academic Press.

Keashly, L. (1998) 'Emotional abuse in the workplace: conceptual and empirical issues', *Journal of Emotional Abuse* 1(1): 85–117.

Keashly, L. and Jagatic, K. (2000) 'The nature, extent and impact of emotional abuse in the workplace: results of a state-wide survey', paper given to the Academy of Management Meeting, Toronto, 4–9 August 2000.

Keashly, L., Trott, V. and MacLean, L.M. (1994) 'Abusive behaviour in the workplace: a preliminary investigation', *Violence and Victims* 9(4): 341–57.

Kets de Vries, M.F.R. (1991) 'The leader's addiction to power', *Journal of Management Studies* 28(4): 339–51.

Kets de Vries, M.F.R. and Miller, D. (1984) *The Neurotic Organization*, San Francisco: Jossey Bass.

Kile, S.M. (1990) *Helsefarleg Leiarskap: Ein Eksplorerande Studie*, Bergen: Rapport til Norges Almenvitenskapelige Forskningsråd.

Kotter, J.P. (1990) *Force for Change: How Leadership Differs from Management*, New York: The Free Press.

Lennane, J. (1996) 'Bullying in medico-legal examination,' in P. McCarthy, M.J. Sheehan and B. Wilkie (eds) *Bullying: From Backyard to Boardroom*, Alexandria, Australia: Millennium Books.

Lewis, D. (1999) 'Workplace bullying – interim findings of a study in further and higher education in Wales', *International Journal of Manpower* 20(1/2): 106–18.

Lewis, D. and Rayner, C. 'Bullying and human resource management: a wolf in sheep's clothing?', paper in preparation.

Leymann, H. (1990) 'Mobbing and psychological terror at workplaces', *Violence and Victims* 5: 119–25.

Leymann, H. (1996) 'The content and development of mobbing at work', *European Journal of Work and Organizational Psychology* 5(2) 165–84.

Liefooghe, A.P.D. (2000) 'Accounts of bullying: beyond individualisation', *Proceedings of the International Harassment Network Annual Conference*, Preston: University of Central Lancashire.

Liefooghe, A.P.D. and Olafsson, R. (1999) '"Scientists" and "amateurs": mapping the bullying domain', *International Journal of Manpower* 20(1/2): 39–49.

Mantell, M. (1994) *Ticking Bombs: Violence in the Workplace*, Burr Ridge, Ill.: Irwin Professional.

Marais, S. and Herman, M. (1997) *Corporate Hyenas at Work*, Pretoria, South Africa: Kagiso Books.

McCarthy, P. (2000) 'The bully-victim at work', paper to Transcending Boundaries Conference, Griffith University, Brisbane, Australia, 6–8 September 2000.

McCarthy, P., Sheehan, M. and Kearns, D. (1995) *Managerial Styles and their Effects on Employees Health and Well-Being in Organisations Undergoing Restructuring*, Report for Worksafe Australia, Brisbane, Griffith University.

McCarthy, P., Sheehan, M.J. and Wilkie, B. (1996) *Bullying: From Backyard to Boardroom*, Alexandria, Australia: Millennium Books.

McGregor, D. (1960) *The Human Side of Enterprise*, New York: McGraw Hill.

Merchant, V. (1997) 'Investigating formal and informal complaints: best practice', The International Harassment Network International Conference on Harassment and Bullying in the Uniformed Services, Burlington Hotel, Birmingham, 29–30 October 1997.

Mullins, L.J. (1998) *Management and Organisational Behaviour* (5th edn) London: Prentice Hall.

Namie, G. and Namie, R. (1999) *Bully Proof Yourself at Work*, Benicia, Calif.: DoubleDoc Press.

Neuman, J. and Baron, R.A. (1998) 'Workplace violence and workplace aggression: evidence concerning specific forms, potential causes, and preferred targets', *Journal of Management* 24: 391–412.

NiCarthy, G., Gottlieb, N. and Coffman, S. (1993) *You Don't Have to Take It! A Woman's Guide to Confronting Emotional Abuse at Work*, Seattle: Seal Press.

Niedl, K. (1996) 'Mobbing and wellbeing: economic and personnel development implications', *European Journal of Work and Organizational Psychology* 5(2): 239–49.

O'Moore, M. (1996) Speech at Trinity College Dublin Conference on Bullying in Schools.

Pavett, C. and Morris, M. (1995) 'Management styles within a multinational corporation: a five country comparison', *Human Relations* 48(10): 1171–91.

Popper, K. (1959 [1934]) *The Logic of Scientific Discovery*, London: Hutchinson.

Power, K.G., Dyson, G.P. and Wozniak, E. (1997) 'Bullying among Scottish young offenders', *Journal of Community and Applied Social Psychology* 7(3): 209–18.

Price Spratlen, L. (1995) 'Interpersonal conflict which includes mistreatment in a university workplace', *Violence and Victims* 10(4): 285–97.

Quine, L. (1999) 'Workplace bullying in an NHS trust', *British Medical Journal* 318: 228–32.

Randall, P. (1996) *Adult Bullying: Perpetrators and Victims*, London: Routledge.

Rayner, C. (2000a) 'The organisational implications of bullying at work', speech to North West Employers Organization, Manchester, 1 March 2000.

Rayner, C. (2000b) 'Building a business case for tackling bullying in the workplace: beyond a basic cost-benefit approach', keynote paper to Transcending Boundaries Conference, Griffith University, Brisbane, Australia, 6–8 September 2000.

Rayner, C. (1999a) 'From research to implementation: finding leverage for prevention', *International Journal of Manpower* 20(1/2): 28–38.

Rayner, C. (1999b) 'A comparison of two methods for identifying targets of workplace bullying', *Abstracts for the Ninth European Congress on Work and Organizational Psychology,* Helsinki: Finnish Institute of Occupational Health, p. 88.

Rayner, C. (1999c) 'Bullying in the workplace', a thesis submitted to the University of Manchester Institute of Science and Technology for the degree of Doctor of Philosophy.

Rayner, C. (1998) 'Workplace bullying: do something!', *The Journal of Occupational Health and Safety – Australia and New Zealand* 14(6): 581–5.

Rayner, C. (1997) 'Incidence of workplace bullying', *Journal of Community and Applied Social Psychology* 7 (3): 199–208.

Rayner, C. and Cooper, C.L. (1997) 'Workplace bullying: myth or reality – can we afford to ignore it?', *Leadership and Organization Development Journal* 18(4): 211–14.

Rayner, C. and Hoel, H. (1997) 'A summary review of literature relating to workplace bullying', *Journal of Community and Applied Social Psychology* 7(3): 181–91.

Rubery, J., Smith, M. and Fagan, C. (1995) *Changing Patterns of Work and Working-time in the European Union and the Impact of Gender Provisions,* Brussels: European Commission.

Savva, C. and Alexandrou, A. (1998) 'The impact of bullying in further and higher education', paper presented to the 1998 Research Update Conference, Staffordshire University, 1 July 1998.

Schuster, B. (1996) 'Rejection, exclusion, and harassment at work and in schools', *European Psychologist* 1: 293–317.

Schein, E.H. (1992) *Organizational Culture and Leadership* (2nd edn), San Francisco: Jossey Bass.

Scott, M.J. and Stradling, S.G. (1994) 'Post-traumatic stress disorder without the trauma', *British Journal of Clinical Psychology* 33: 71–4.

Senge, P., Roberts, C., Ross, R., Smith, B., Roth, G. and Kleiner, A. (1999) *The Dance of Change,* New York: Doubleday.

Sheehan, M. (1999) 'Workplace bullying: responding with some emotional intelligence', *International Journal of Manpower* 20(1/2): 57–69.

Sheehan, M. (1996) 'Case studies in organisational restructuring', in P. McCarthy, M. J. Sheehan and B. Wilkie (eds) *Bullying: From Backyard to Boardroom,* Alexandria, Australia: Millenium Books.

Smith, P.K., Singer, M., Hoel, H. and Cooper, C.L. (2002) 'School bullying and workplace bullying: are there any links?', paper under submission.

Sutton, J. (1998) 'Bullying: social inadequacy or skilled manipulation?', PhD thesis submitted to Goldsmiths College, Psychology Department, London.

Tattum, D. and Tattum, E. (1996) 'Bullying: a whole school response', in P. McCarthy, M. J. Sheehan and B. Wilkie (eds) *Bullying: From Backyard to Boardroom,* Alexandria, Australia: Millenium Books, pp. 13–23.

Thylefors, I. (1987) 'Syndbockar. Om utstøtning och mobbning i arbeidslivet', *Natur och Kultur*, Stockholm.

Tjosvold, D. (1995) 'Effects of power to reward and punish in co-operative and competitive contexts', *Journal of Social Psychology* 135(6): 723–736.

TUC (1997) 'Hard times: 5,000 reasons for new rights at work. Summary and analysis of calls to the TUC's bad bosses hotline', Trades Union Congress, Campaign and Communications Department, London.

UNISON (2000) *Police Staff Bullying Report* (number 1777), London: UNISON.

UNISON (1997) *UNISON Members' Experience of Bullying at Work*, London: UNISON.

Van de Vliert, E. and de Dreu, C. (1994) 'Optimizing performance by conflict stimulation', *International Journal of Conflict Management* 5(3): 211–22.

Van de Vliert, E., Euwema, M.C. and Huismans, S.E. (1995) 'Managing conflict with a subordinate or a superior: effectiveness of conglomerate behavior', *Journal of Applied Psychology* 80(2): 271–81.

Vartia, M. (1996) 'The sources of bullying: psychological work environment and organizational climate', *European Journal of Work and Organizational Psychology* 5(2): 203–14.

Verbrugge, L.M. (1985) 'Gender and health: an update on gender and evidence', *Journal of Health and Social Behaviour* 26: 156–82.

Withey, M. and Cooper, W. (1989) 'Predicting exit, voice, loyalty and neglect', *Administrative Science Quarterly* 34: 521–39.

Wilkie, B. (1996) 'Understanding the behaviour of victimised people', in P. McCarthy, M. J. Sheehan and B. Wilkie (eds), *Bullying: From Backyard to Boardroom*, Alexandria, Australia: Millennium Books.

Witheridge, L. (1998) Personal communication to C. Rayner.

Worrall, L. and Cooper, C.L. (1999) *The Quality of Working Life: 1999 Survey of Managers' Changing Experiences*, London: The Institute of Management.

Yukl, G. (1994) *Leadership in Organizations* (3rd edn), Englewood Cliffs, NJ: Prentice Hall.

Yukl, G. and Falbe, C.M. (1991) 'Importance of different power sources in downward and lateral relations', *Journal of Applied Psychology* 76(3): 416–23.

Zapf, D., Knorz, C. and Kulla, M. (1996) 'On the relationship between mobbing factors, job content, social work environment, and health outcomes', *European Journal of Work and Organizational Psychology* 5(2): 215–37.

Index